What We Inherit

HOW NEW TECHNOLOGIES
AND OLD MYTHS ARE SHAPING
OUR GENOMIC FUTURE

SAM TREJO

DAPHNE O. MARTSCHENKO

PRINCETON UNIVERSITY PRESS
PRINCETON & OXFORD

Graphics courtesy of Dayan D'Aniello.

Illustrations courtesy of Allegra Martschenko.

Published by Princeton University Press
41 William Street, Princeton, New Jersey 08540
99 Banbury Road, Oxford OX2 6JX

press.princeton.edu

GPSR Authorized Representative: Easy Access System Europe - Mustamäe tee 50, 10621 Tallinn, Estonia, gpsr.requests@easproject.com

ISBN 9780691237756
ISBN (e-book) 9780691237749

Library of Congress Control Number: 2025937714

British Library Cataloging-in-Publication Data is available

Editorial: Rachael Levay and Tara Dugan
Production Editorial: Theresa Liu
Jacket/Cover Design: Haley Jin Mee Chung
Production: Erin Suydam
Publicity: William Pagdatoon
Copyeditor: Karen Verde

Jacket image: © Dayan D'Aniello / Day& Co.

This book has been composed in Arno and Helvetica Neue LT STD

Printed in the United States of America

10 9 8 7 6 5 4 3 2 1

For our favorite companions: Ali, Marissa, Pickle, and Sassy

CONTENTS

PREFACE

DAPHNE: *I think I first became interested in genetic myths in autumn of 2016. I was a PhD student in Education at the University of Cambridge and had just returned to the United States to begin collecting data. For nearly six months, I observed and interviewed teachers at two vastly different schools. The first was a private school that provided tailored services to children identified by psychologists as gifted. The second was a public charter school that had a lottery-based admission system. The private school was located 30 miles outside of Chicago; it had a mostly White and Asian student body, and its annual tuition was $19,000. The charter school was located on Chicago's west side; its students were mostly Black and working-class.*

During those six months, I noticed some striking similarities in the attitudes of the teachers I was interviewing—even though they worked in starkly different school contexts. I remember one of the teachers at the private school describing what he thought made his students different from those in Chicago public schools. He said that a child's home environment "has a part to do with it," but he also emphasized that "some people are ahead in the game—some people have those great, great genes." Just a few weeks later, a teacher at the charter school remarked to me that she felt every student has "a different amount of whatever natural gifting." In both schools, teachers believed that DNA offered at least a partial explanation for why some students perform better than others in the classroom.

It made me wonder about the consequences of this kind of thinking: how a teacher's beliefs about their students impacted those students' beliefs about themselves.

While it wasn't until I began work on my PhD that I became interested in the effects of genetic myths specifically, I've been curious about the extraordinary power of stories for as long as I can remember. As a kid, I loved reading science fiction and fantasy books; as I grew older, I turned to memoirs and narrative nonfiction. Ultimately, I think my interest in stories led me to a career as a bioethicist at Stanford University. I spend my time exploring the sociocultural processes that underlay how we see ourselves and others; these processes, and the ugly history behind them, are key for thinking about the ethical and social implications of emerging genomic technologies. I'm driven to find answers to questions like: How do myths about genes shape collective understandings of biological inheritance and human difference, and what role do these myths play in the reproduction of inequality today?

SAM: *The fall of 2016 was a pivotal career moment for both of us, wasn't it? Because that was when I first became interested in the effects of DNA. I was just twenty-two years old—a first-year PhD student studying a unique blend of sociology and economics. I'd recently arrived at Stanford University with just the few possessions that I could cram into my '99 Toyota minivan, as well as a keen interest in statistics and social inequality.*

A few weeks into my PhD, a professor in my department invited me to the semester's first meeting of an informal research group. There, I met a motley crew of Stanford researchers who hailed from fields like sociology, economics, epidemiology, psychiatry, biology, and genetics. The group became known as the Genomics and Social Science journal club, or "G&SS" for short.[1] We gathered to discuss and debate the seemingly endless supply of genomic studies coming out in scientific journals each week—especially ones that focused on human behavior. Many of these new studies were applying the same scientific tools used to study clinical traits, like blood pressure and Crohn's disease, to instead study how far people go in school or whether they smoke cigarettes.

G&SS met monthly on Fridays in the late afternoon. The wine and conversation flowed freely, and my time there played a defining role in my graduate school experience. The group served as my introduction to both human molecular genomics and the unique power of DNA to stoke controversy. Our discussions often centered on two questions that I continue to ask today as a sociologist at Princeton University: How does the DNA that a person inherits help shape who they eventually become, and, what the heck should we make of these new things called polygenic scores?

DAPHNE: *I still remember the day we first met. . . . We each began our respective journeys toward writing this book in 2016, but our paths didn't actually cross until the following year. We were attending the same education conference in New York City and delved into our shared interest in DNA, ethics, and social inequality over a massive lunch at a Chinese restaurant. Those early conversations fooled me into thinking our disagreements would be simple and straightforward: I would be the critic of new genetic technologies and you, the enthusiast.*

SAM: *Right, like fire and ice. I'm a social scientist who utilizes DNA in my research. You're a bioethicist who examines the benefits and harms of that research. I study how the DNA that a person inherits impacts their life, while you focus on how the wrongheaded beliefs about genes that we as a society still hold affect people in material ways.*

DAPHNE: *Well, we learned pretty quickly that first-impression disputes are rarely as simple as they first seem.*

SAM: *True. Despite our surface-level differences—or even opposition!—we were both curious about the risks and benefits of different types of scientific research. We also shared a feeling of frustration regarding the bitter and entrenched academic debate about genomic technologies.*

DAPHNE: *Yeah, and although talking through these issues isn't always easy, I think we both feel it's urgent that we do so. While researchers argue in the ivory tower, for-profit companies are starting to find avenues for selling new, often unvalidated and inaccurate DNA-based tools like polygenic scores to everyday consumers.*

SAM: *Many people may not realize it, but we're in the midst of a unique historical moment when it comes to DNA. Thinkers have been interested in the biological transmission between parents and their children for thousands of years—since at least Plato and the Greeks. And, while scholars have theorized about family resemblance for a long time, it's only in the last century that we figured out which specific molecule gets passed on and what it looks like chemically. It's only been in the last decade that we've been able to collect enough DNA to make rigorous discoveries. So, there are these big questions about family and biology that have been waiting for millennia for an answer, and it seems like the data we need is finally at our fingertips.*

DAPHNE: *True, but as society works with and processes this new information, it also kicks up myths and controversies about genetic difference. I worry most about this aspect—the damaging myths about DNA that have swirled around for hundreds of years. These genetic myths were created before we even had access to DNA, but now I see people arguing—incorrectly, I might add—that genomics research is providing evidence of their legitimacy. What concerns me is that the molecule inevitably seems to come with the myth.*

———

This book is an adaptation of *adversarial collaboration*, a style of research that emerged in the early 2000s to try to make disagreement more productive.[2] The goal of an adversarial collaboration is not necessarily to arrive at a consensus, as the pressure to achieve consensus can deter conversation and collaboration between "adversaries." Instead, an adversarial collaboration sets out to identify key disagreements between collaborators and understand the roots of those disagreements.[3] The approach began in psychology and behavioral economics, first coined by the Nobel laureate Daniel Kahneman. After decades in academia, Kahneman was tired of "angry science"—researchers publishing scathing critiques of each other's work, back and forth, again and again, with almost no direct conversation happening in good faith. He wanted to find a way to make the empirical disagreements in his research area more productive. So, rather than avoid his "adversaries," Kahneman sought them out.

Soon after we first met in 2017, the kernel of an idea began to form: What if the two of us came together to try to understand each other and our differences? Would we learn anything in the process that could help consumers, policymakers, and our academic colleagues navigate the growing onslaught of DNA-based tools and applications? This book is the culmination of years of often-heated conversations in which we tried to answer these questions.

When we embarked on this project, we imagined a book that would largely focus on our disagreements as two researchers working in distinctly different fields and using distinctly different methods: Sam, the quantitative sociologist who leveraged genomic data, and Daphne, the qualitative bioethicist who critiqued it. However, *What We Inherit* turned out quite differently from that original vision. Kahneman said that his adversarial collaborations ended with "new facts accepted by all, narrowed differences of opinion, and considerable mutual respect."[4] Our experience writing this book mirrors Kahneman's. As a result, and to our surprise, this book mostly takes on one, united voice.

Part 2, though, reveals that there are still plenty of disagreements that require us to separate and distinguish between our unique and opposing perspectives. To capture these differences, we adopt a dialogue format at the start and end of chapters 4, 5, and 6. (Like the opening section of this preface, these sections will be distinguished typographically from the main portions of the book.)

A number of books have been written about social genomics,[5] but none have brought together a bioethicist and quantitative social scientist. Our distinct—and sometimes entrenched—academic communities rarely meaningfully engage with one another. In fact, both of us were advised by colleagues *not* to embark on such a project. Our very willingness to work together implies that we may be more moderate in our views than our broader respective communities. Thus, our work resides in what others who have done adversarial collaborations call the "murky middle."[6] Importantly, our goal with this book is not to fully represent both of our communities; rather, we aim to provide an example of constructive conversation about controversial issues that avoids "straw manning" either side of the debate. We view this book as a first step in bridging a divide and hope that others, inside and outside of academia, might pick up where we leave off. *What We Inherit* provided us with an unexpected antidote to entrenchment. Despite our differences, writing this book has solidified our shared belief that *to progress debates about difficult and charged topics and to ensure that scientific discoveries benefit all and not just some, people need to take some time to listen—really listen—to each other.*

Through the process of questioning ideas that often seem obvious and firmly ingrained to researchers, the opposition and conflict between us gradually neutralized. In the end, the "other side" seemed more like a challenging ally than an adversary. Perhaps most importantly, our book demonstrates that developing a regulatory framework for safeguarding against many of the risks presented by DNA-based tools like polygenic scores does not require complete consensus. Conversation is urgently necessary in the face of unregulated uses of modern genetic technologies. We hope our collaboration takes an important step toward bringing together communities that tend to talk past, rather than with, one another.

ACKNOWLEDGMENTS

Every word of this book was written collaboratively, with the two of us as equal contributors. As our first acknowledgment, we have to thank each other. Somehow, a friendship grew out of a lot of arguing. There are countless other folks who helped make this book a reality—it truly took a village. It would be impossible to name everyone who provided support, guidance, and mentorship throughout the book-writing process, but we will nevertheless give it the old college try.

What We Inherit would have never materialized if Ben Domingue hadn't introduced two wide-eyed and naïve PhD students to each other back in 2017. We would have also never finished what we started a few years later without our diligent copyeditor and writing assistant, Emma Daugherty. And, it was the creativity of Dayan D'Aniello and Allegra Martschenko that made the book's beautiful graphics and illustrations possible. Allegra also served as our all-around guru on what was to us a very new and unfamiliar publishing process. We are so grateful to all of you.

We would also like to thank Mildred Cho, Dalton Conley, Jason Fletcher, and Paige Harden, who all provided crucial guidance and support back when the possibility of writing a book together was merely an idea. We are deeply grateful to Erik Parens for suggesting that we apply to the Brocher Fellowship and to the Fondation Brocher for accepting us and providing us with time to write in peace in one of the most beautiful places on the planet. We are also indebted to the many kind and supportive colleagues who comprise each of our academic homes—Princeton University's

Department of Sociology and Stanford University's Center for Biomedical Ethics. In addition, we must acknowledge all the help we received from our beloved friends and family members. Thanks to our parents Alex, Toyin, Steve, and Nina and to our siblings Allegra, Alexandra, Diana, and Max. We also owe a debt of gratitude to Farhana Ferdous, Alexandra Martschenko, Alex Soble, Marissa Thompson, and Stephen Trejo—who read and provided feedback on the very first draft of our book.

We are also grateful to our many colleagues, including Jedidiah Carlson, Dalton Conley, Paige Harden, Anna Lewis, Iain Mathieson, Lucas Matthews, Molly Przeworski, Wendy Roth, Maya Sabatello, Matt Salganik, Jim Tabery, and Patrick Turley—as well as two anonymous peer reviewers—who carefully and thoughtfully read portions of the book and lent us their expertise. Special thanks to Courtney Hoggard, who provided research support in the early stages of the project. We also wish to acknowledge the many folks who listened to us give academic talks, both jointly and independently, on some of the ideas presented in this book. This group of wonderful listeners includes but is not limited to audiences at Columbia University, the University of California, Berkeley, the University of California, Santa Cruz, the University of Colorado, Boulder, the University of Minnesota, the University of Oslo, the University of Oxford, the University of Pennsylvania, the University of Southern California, the University of Texas at Austin, the Centers for Excellence in ELSI Research, and the European Social Science Genomic Network.

Finally, if Princeton University Press hadn't taken a chance on the two of us, this book would never have found its way into print. Many thanks to our editor, Rachael Levay, and her editorial assistant, Erik Beranek. Rachael, thank you for your thoughtful guidance, unending patience, and earnest enthusiasm! And, of course, we owe so much to our spouses, Ali and Marissa. You two spent the better part of six years hearing us complain about the challenges of writing this book—and sometimes about each other. Thank you for keeping us sane. We love you both, the lives we have with you, and our four-legged companions, Pickle and Sassy.

PART I
Myths and Molecules

1

Two Acids

In the 1960s, a Harvard psychologist named Robert Rosenthal and an elementary school principal named Lenore Jacobson teamed up to conduct an unusual (and arguably unethical) research study.[1] They hoped to complicate theories about why certain students succeed academically, which at the time tended to center on the psychological characteristics of a child, like personality and intellect. Rosenthal and Jacobson, however, wondered about the importance of how adults *perceive* a child, irrespective of the child's actual characteristics. This question proved a classic case of "the chicken or the egg"; the cause and effect seemed impossible to disentangle. Do students succeed because their teachers believe in them, or do teachers form their beliefs about students based on characteristics that lead to success? The duo devised an experiment to settle the matter: Using Jacobson's elementary school in South San Francisco as a laboratory, the pair rounded up 300 first- and second graders and administered the "Harvard Test of Inflected Acquisition." When the test results came back, Rosenthal and Jacobson gave each teacher a list of the students in their classroom who were "bloomers"—or children whose results indicated that they would likely excel academically in the coming years.

Unbeknownst to the teachers, the Harvard Test of Inflected Acquisition didn't actually exist! Instead, Rosenthal and Jacobson had administered a common IQ test. Either way, the test didn't matter: Rosenthal and Jacobson had randomly selected the so-called bloomers, meaning that they were no different than any of the other students. A year later, Rosenthal and Jacobson discovered that the bloomers had learned at faster rates than their peers. Teachers' beliefs about which students had inherent potential for academic success had become a self-fulfilling prophecy. While Rosenthal and Jacobson's experiment is the subject of contentious scientific debate (their sample size was small, and they have been accused of cherry-picking the statistical results), recent studies using more rigorous methods confirm that teachers' expectations really do impact their students, even if the effects may be more modest than Rosenthal and Jacobson originally argued.[2]

Rosenthal and Jacobson's experiment pushed researchers to break away from an outdated paradigm that focused only on a child's mind, shifting the perspective to also consider the child as a social object molded by external influences. The study demonstrated that ideas and perceptions, even unfounded ones, have the power to shape a person's life trajectory. Like Rosenthal and Jacobson's experiment, this book illustrates the ways in which ideas and perceptions, however untrue, shape people's understandings of themselves and others. Our goal is to expand the way in which people think about genes, broadly conceived. In particular, *What We Inherit* outlines two intertwined inheritance processes: DNA itself, and the myths about genes that also span generations.

You may think of genes in terms of DNA or deoxyribonucleic acid: a molecule that sits in the center of a cell and acts as a kind of biological instruction book. DNA plays a key role in the evolution and adaptation of a wide range of life-forms, from fungi to palm trees to whales—and, of course, human beings. Furthermore, genes function as an iconic social object with a powerful grip on the human collective imagination.[3] The literal acid within cells gets passed down biologically from parent to child; the conceptual acid, or stories and myths about genes and how they affect human life, gets passed down culturally from a eugenic past.

For better or for worse, the train has left the station; novel genomic technologies and discoveries have already begun to accumulate and disseminate throughout a range of life domains—from academic research to the direct-to-consumer genetic testing industry to the fertility clinic. The growing importance of DNA demands new frameworks for understanding human genetic differences and for considering the regulation of genomic tools. *What We Inherit* argues that a full account of the power and influence of genes must consider the dual inheritance processes of DNA and genetic myths. To ensure that the benefits of the unfolding genomic era are maximized and its risks minimized, researchers and policymakers need to account for historical and social context when deciding what research to conduct and how to discuss it. Careful attention must be paid to social inequalities, past and present, when considering how new genomic technologies may be used in healthcare, schools, industry, and society writ large. This book stems from a shared motivation to combine expertise in two acids that society and its members inherit— Daphne, the myth, and Sam, the molecule.

Better understanding the acids that human beings inherit requires rewinding to the birth of the modern field of human genomics. Its genesis at the turn of the twenty-first century resulted from a high-profile and dramatic clash, the likes of which the often-mundane world of scientific research rarely sees. The field's contentious birth would foreshadow the many controversies surrounding DNA to come in the following decades.

The Human Genome Project launched in 1990 with an ambitious and unprecedented scientific goal: to map out the entire DNA sequence of the human species, from beginning to end. Led by Francis Collins, the director of the National Human Genome Research Institute (the primary government-funded genomics institute in the United States), the Human Genome Project intended to make its discoveries widely and freely available to researchers around the world. To the frustration of Collins and his team, Craig Venter, a researcher-turned-entrepreneur who founded a company called Celera Genomics, had a different plan; he hoped to privately sequence the genome and sell the ensuing scientific discoveries.

Each group raced to finish their sequencing first, and the competition quickly turned nasty. Venter publicly criticized the Human Genome Project, calling it a waste of public resources. One of the Human Genome Project's leading scientists shot back, calling Celera's commercialization of the genome a "con-job."[4] Eventually, the White House stepped in and initiated peace talks between the two groups of researchers. It took time (and pizza), but ultimately Venter and Collins were able to resolve their dispute. In 2000, when researchers finished a first draft of the human genome, Collins and Venter both flanked President Bill Clinton as he announced: "Today we are learning the language in which God created life . . . With this profound new knowledge, humankind is on the verge of gaining immense new power to heal."[5]

The Human Genome Project wrapped up in 2003, two years ahead of schedule. Some say the rapid technological changes following the completion of the project amount to a "DNA revolution." Over the last two decades, the cost of DNA sequencing has dramatically decreased. It took thirteen years and cost nearly $3 *billion* to sequence the very first human genome; today, it costs only a few hundred dollars to sequence a genome in less than 24 hours.

What exactly is DNA—the molecule that the Human Genome Project thrust into view? Think of a person's DNA as a figurative "book" of biological instructions written using an alphabet of four letters: adenine (A), cytosine (C), guanine (G), and thymine (T). Each letter, or nucleobase, is paired with a complementary letter to form base pairs. A pairs with T, and C pairs with G. Each human has roughly three billion base pairs comprising their DNA sequence, which are organized into forty-six "chapters" (or *chromosomes*). The ordering of As, Cs, Ts, and Gs is, for the most part, the same for all humans. That is, every human being has a very similar book of DNA; nonetheless, each person's DNA sequence is one of a kind, almost like a fingerprint. (The exception to this rule, of course, is identical twins.) Each person inherits a unique DNA sequence from their parents, and that sequence remains unchanged throughout a person's entire lifetime. When one person has a certain letter (or set of letters) at a particular location in the genome (say a T), whereas another person has a different letter at that same spot (say a G), those two people have different *DNA variants*.

In the decades since the completion of the Human Genome Project, scientific advances in collecting and analyzing genomic data have produced a torrent of discoveries linking DNA to a wide range of human traits. For thousands of years, the ability to understand the effects of DNA hinged on observing it indirectly via familial relatedness (for example, by comparing identical and fraternal twins, or siblings and cousins). Now, it is possible to observe each person's unique ordering of *A*s, *C*s, *T*s, and *G*s at the molecular level. Can an individual's DNA data be used to make predictions about their life outcomes—for instance, which diseases they will come to develop or how their personality will change as they age?

If you took high school biology in the United States, you may remember learning about an Austrian monk named Gregor Mendel who conducted experiments with peas. Mendel cross-pollinated different kinds of pea plants (crossing tall plants with short plants and yellow plants with green plants, for example) to try to understand how traits get passed down between generations. He is credited with discovering many of the basic principles of genetic inheritance—namely, that organisms pass portions of their DNA to their offspring (although the term "DNA" hadn't yet been coined in Mendel's day). Mendel argued that, for every trait, offspring inherited one DNA variant from each parent. Some of these variants, he concluded via experimentation, are dominant, while others are recessive.

Until the DNA revolution, most researchers believed that just a single or a few DNA variants impacted a given trait; traits influenced by a single region of the genome known as *monogenic*. Huntington's disease, for example, is a monogenic trait caused by a DNA variant of the HTT gene on chromosome 4. Similarly, sickle cell anemia is a monogenic trait caused by a DNA variant of the β-globin gene on chromosome 11. High school classes in the United States rely on Mendel's experiments and monogenic traits to introduce students to genetics (remember Punnett squares?). In an abstract classroom environment, DNA may first seem to operate in a clear-cut and straightforward manner; however, the technological advancements brought on by the completion of the Human Genome Project paint a markedly different picture.

A key discovery of the genomic era is that most human characteristics are *not* monogenic; instead, they are *polygenic*, or complex. You can throw "dominant" and "recessive" out the window; polygenic traits, like height, are influenced by countless DNA variants dispersed widely across the genome. There is no "height gene"—a lone variant responsible for genetic influences on how tall a person grows to be. Instead, thousands (or even *millions*) of DNA variants are correlated with a person's height.[6] For a polygenic trait, any given DNA variant contributes just a tiny fraction of the total genetic influence.

As a method of summarizing these myriad DNA variants, researchers have developed a new genomic tool known as a *polygenic score*. (Note that polygenic scores are also referred to as polygenic indexes,[7] polygenic risk scores, and genetic risk scores.) Polygenic scores use a person's DNA to make predictions for a wide range of outcomes—for example, how tall they will likely grow, their chances of developing skin cancer, and what level of education they will reach. While there are very few traits that are "genetic" in a monogenic sense, the vast majority of traits are "genetic" in a polygenic sense. The realization that most traits are associated with many DNA variants rather than one or just a few has transformed the focus of human genomics from specific inherited diseases to almost all types of individual difference.

As the costs of collecting and analyzing DNA continue to drop and genomic databases grow in size, the predictiveness of polygenic scores also continues to improve. Importantly, the predictions offered by polygenic scores are probabilistic, not deterministic. To visualize the probabilistic relationship between a polygenic score and an outcome (like educational attainment), researchers use scatterplots; figure 1 displays three. Each small grey dot in figure 1 represents a single person, and all three plots contain data on the same 874 American adults. These 874 Americans were all born around 1980 and are participants in the National Longitudinal Study of Adolescent to Adult Health, an ongoing biosocial survey used by researchers studying health and human behavior.[8] The variable on the vertical axis is each person's educational attainment—the number of years of formal schooling each individual completed. For instance, those who dropped out of high school without graduating have fewer years of schooling than those who got their high school diploma. Those who entered the workforce after high school graduation, in turn, have fewer years of schooling than those who continued on to university. The variables on the horizontal axes are three different polygenic scores. Each person's polygenic score is generated by statistically combining their unique string of As, Cs, Gs, and Ts to produce a single number, ranging from about -3 to 3, that will correlate with their eventual educational attainment.

The punchline of figure 1 is that the predictive accuracy of polygenic scores has increased significantly over the past two decades. The DNA sequences of the 874 Americans represented in figure 1 are identical across the three panels (and so is their educational attainment). What changes across the panels is the precise formula used to transform someone's DNA sequence into a polygenic score. A person's polygenic score for a given trait is essentially an enormous weighted average of their genome, where the weights represent researchers' best guess of which DNA variants are associated with increases or decreases in the trait in question (and by how much). In 2004, at the start of the genomic era, it was virtually impossible to make a prediction about someone's educational attainment using DNA, and when the first major genomic study of educational attainment was published in 2013, polygenic prediction was just barely possible.[9] Now, almost a decade later, the formula underlying the educational attainment polygenic score has substantially improved. The latest version of the score is just as predictive of eventual years of schooling as all the textbook variables used by social scientists: a child's family income, their IQ test scores,[10] and their parents' education levels. (In nationally representative data sources, this polygenic score explains about one-sixth of the variation in years of schooling between individuals.)[11]

Figure 1 also highlights the quite limited ability of a polygenic score (or any variable, for that matter) to predict a specific person's educational attainment. Notice the often-large vertical gap between the dark black line, which represents a person's predicted education (given their polygenic score), and the gray dots, which represent a person's realized education. Even the best predictors leave the vast majority of variation unexplained. Plenty of people with high polygenic scores for educational attainment do not graduate high school, and plenty of people with low polygenic scores for educational attainment end up graduating from college.

Still, perhaps because of internalized genetic myths and biases, people can often erroneously think that the information gleaned from a polygenic score is as definitive as learning they are a carrier for Huntington's disease. They may also believe that polygenic scores have managed to somehow disentangle the effects of a person's DNA from their environmental context. Concerns over what exactly a polygenic score captures and how they should be used continue to mount. This book shows just how complicated these so-called complex traits (and the polygenic scores that try to predict them) really are and discusses how to navigate these complexities.

Polygenic scores are making swift inroads into society, leaving some excited and others on edge. Hoping to build a world in which more people experience better health and social outcomes,

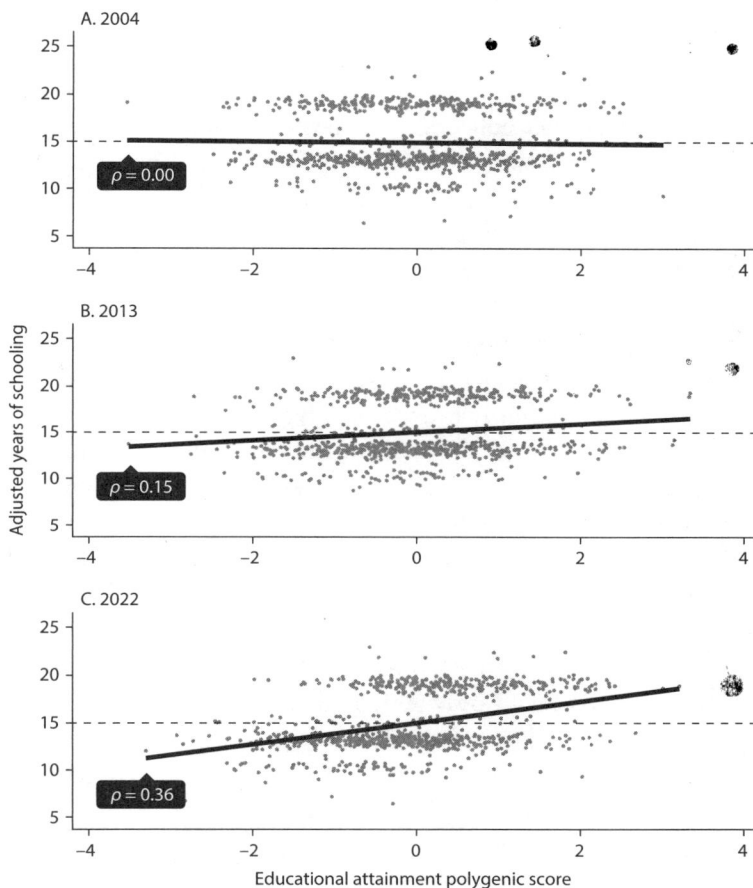

FIGURE 1. Polygenic Prediction of Educational Attainment: 2004 through 2023. Each panel of the figure displays a scatterplot containing the same 874 individuals of European ancestries from the National Longitudinal Study of Adolescent to Adult Health. Years of schooling is measured using the International Standard Classification of Education (ISCED) 1997 classification system and is collected at Wave IV of the study. Years of schooling is statistically adjusted for sex, age, and ten genomic principal components.[12] The same years of schooling variable is used in all three panels, but each panel plots a different polygenic score variable. Panel A utilizes a randomly generated standard normal variable, representing polygenic prediction during the candidate gene era. Panel B utilizes a polygenic score generated from the results of the very first genome-wide association study (GWAS) of educational attainment ($N = 126{,}559$ individuals[13]). Panel C utilizes a polygenic score from the most recent GWAS of educational attainment ($N = 3{,}037{,}499$ individuals[14]).

proponents of polygenic scores believe these new tools have the potential to advance collective understanding of how well certain interventions or treatments work (and for whom). They see polygenic scores contributing to the ongoing projects of mapping the complexities of being human and improving people's well-being. Still, the United States has a long and fraught history of connecting DNA to human behavior. For this reason, polygenic scores for social and behavioral traits like educational attainment carry a particular historical baggage and set of social risks; they are sometimes viewed differently from, say, polygenic scores for diseases like breast cancer.[15] While many of the arguments made in this book apply to a broad range of genetics research, an understanding and acknowledgment of this ugly history inspires a particular focus on *social genomics*, a research field that seeks to connect a person's DNA sequence to social and behavioral traits, like a person's sexual orientation or occupation. Such research is at the highest risk of perpetuating the damaging types of myths described in this book.[16]

Rapid increases in polygenic prediction, coupled with the growing number of traits for which polygenic scores are available, raise a range of questions: What sort of information do polygenic scores provide? What are their risks and benefits? How should new DNA-based tools be regulated in society? As debates over these questions rage on, polygenic scores are increasingly being used by companies and institutions with surprisingly little oversight. Society needs to brace for a future where polygenic prediction is readily available. Now is the time to start having tough conversations about how to best navigate society's bumpy landing into this new genomic era.

———

Genes—whether through the literal acid of DNA or the conceptual acid of genetic myths—are sources of polarization and pride. They comprise what the physician and writer Siddhartha Mukherjee calls "one of the most powerful and dangerous ideas in the history of science."[17] DNA provides the fundamental means by which all humans differ from one another. Layered over variation in people's individual DNA profiles are the social myths created, narrated, and passed down through generations. These myths frame a collective understanding of how genes function and what they say about who people are.

This book pushes forward a much-needed discussion about how society should and should not use genetic information like polygenic scores. Today, people with a few hundred dollars to spare can spit into a tube or swab their cheeks using an at-home test; they'll receive information on their polygenic scores for traits that range from Type 2 diabetes and prostate cancer to math ability and intelligence. Increasing access to these new genomic technologies presents both opportunities and challenges, many of which inflect age-old debates about genetic difference. Real-world applications of polygenic scores are growing, but regulation lags behind—in part because of how sensitive and charged conversations about DNA can be.

At the same time, influential genetic myths about DNA are shaping people's perceptions of polygenic scores. These myths threaten to grow in power and influence due to the ever-increasing presence of genomic research and genomic tools. The two specific genetic myths unpacked in this book—the *Destiny Myth* and the *Race Myth*—are both socially inherited from historical eras marred by the powerful legacies of eugenics and scientific racism.[18] These myths have been passed down through books and laws, folklore and oral tradition. This book aims to pull them apart, bit by bit, until they are exposed for what they truly are: fictions that serve to distract from real issues like racial and class-based social inequalities,[19] the unsavory realities of an unequal world.

What We Inherit consists of three parts, each of which includes three chapters. Part one (chapters 1–3) provides background and context on the history of genomic research and genetic myths. Chapter 2 helps to put the infamous "nature vs. nurture" debate to bed by explaining what it means for DNA to influence a person's health, behavior, or life outcomes. The environment and DNA interact in complex ways, and phrases like "genetic effects" or the "effects of DNA" can mistakenly lead to overstatements of the role of DNA. The chapter concludes by debunking the Destiny Myth: the flawed idea that **the effects of DNA are immutable and inevitable**. (Note that causal relationships between DNA and outcomes are often referred to as "genetic effects," but this book uses the term the "effects of DNA" because this terminology choice helps to distinguish them from the effects of genetic myths.)

Chapter 3 disentangles race and ancestry. Race is a sociopolitical construct designed to benefit some and harm others; it differs from the ancestral information captured by the large Family Tree of humanity. In the twenty-first century, direct-to-consumer genetic testing companies can provide information about where a person's ancestors may have originated, but that information needs interpretation through an informed lens (and with a cautionary grain of salt). In disentangling race and ancestry, the chapter reveals the lies behind the Race Myth: the false belief that **DNA differences divide humans into discrete and biologically distinct racial groups**.

In a perfect world, half of the published copies would've listed Daphne first on the cover, and the other half would've listed Sam first. (In our actual world, we flipped a coin and Sam won.) However, while the book largely takes on one, united voice, the authors do not agree on everything. Their disagreements come through particularly in part two (chapters 4–6), which focuses on debates regarding genetic myths and genomic research. Chapter 4 explores the historical impacts of genetic myths on society, as well as the persistence of these myths. Chapter 5 covers debates about the promises and pitfalls of social genomics and discusses who gets to weigh the risks and benefits of such research. This chapter also examines how genetic myths can repurpose modern-day genomic research. Chapter 6 explores disagreements about whether DNA is relevant for understanding and ameliorating social inequality; at the heart of this debate are different understandings and definitions of social inequality. This chapter provides a conceptual basis for thinking about the regulation of polygenic scores (covered in part 3). As noted in the preface, in order to clarify and illustrate our disagreements, each chapter in part 2 begins and ends with dialogues between Sam and Daphne: for instance, on how genetic myths impact American society, the risks and benefits of social genomic research, and whether DNA "matters" when considering social inequality.

Part 3 provides policy recommendations for navigating a modern world rife with polygenic scores, focusing on three specific applications of the DNA-based tool. Chapter 7 considers the regulation of polygenic embryo selection, a technology that allows prospective parents to choose certain genetic characteristics of their future children. Utilizing existing polygenic scores in the fertility clinic can produce increases in height of 2 ½ inches, but the technology is expensive and—at present—not very effective for most people. Chapter 8 discusses the regulation of direct-to-consumer genetic testing and polygenic-informed screening programs. As online genetic tests proliferate and as polygenic scores are beginning to stratify care in hospitals and clinics, new frameworks must guide the ethical and responsible use of genomic information. In both chapters, the goal of the policy recommendations is to prevent these polygenic scores applications from, at the very least, widening preexisting structural inequalities. Chapter 9 presents concluding remarks and key takeaways, offering next steps for researchers, policymakers, and members of the public.

Now more than ever, reaching across the aisle and having conversations with those you disagree with may feel fraught and unproductive. This book aims to convince you that many such debates are necessary—and even urgent. Society faces a welter of big, challenging questions: Is it possible to balance social equality and efficiency in the face of rapid technological changes? What might it look like to conduct scientific research in a way that delivers vital discoveries without repeating past mistakes? To what extent is it appropriate to shape the biology of future generations of humans? Creating a world in which genomic data is used in a socially responsible way requires that these questions be answered.

As you will see, however, there are plenty of areas where the authors disagree with one another. Writing this book reflects a shared commitment to the belief that, even as scientific advances increase researchers' abilities to zoom in and view processes at the molecular level, efforts to also seek macro-level explanations of human processes cannot fall by the wayside. Working together, the authors slowly begin to digest the rapid development of new genomic technological changes and the age-old genetic myths that accompany them; along the way, this book aims to humanize the people affected by both. Navigating new genomic technologies while putting to rest old genetic myths is not going to be easy, but the only path forward is to learn how to listen and talk *with* rather than *past* each other.

2

The Destiny Myth

The mid-2000s was a flashy era of pop stars and boy bands, low-rise jeans and frosted tips. Against this baffling cultural backdrop, Amanda Wanklin and Michael Biggs fell in love. Amanda worked as a healthcare aid, and Michael ran an auto-repair business. The couple decided to settle down in Birmingham, England, a city known for its manufacturing roots and a renowned rock music scene (which launched bands like Black Sabbath and Duran Duran). Amanda and Michael were eager to start a family; as an interracial couple, they knew that not everyone would approve of their union, but Amanda and Michael simply "didn't give a toss."[1]

At first, Amanda and Michael struggled to conceive a child. Undeterred, they decided to utilize in vitro fertilization (IVF); they were elated when Amanda became pregnant with twins. On a warm day in July 2006, Millie Marcia Biggs and Marcia Millie Biggs were born by Caesarean section. Though twins, Millie and Marcia were decidedly *not* identical. Even from a young age, the sisters' differences in physical appearance were striking. Marcia, with pale skin, blue eyes, and dirty blonde hair, takes after her mother. Millie, on the other hand, has brown eyes, dark hair, and caramel skin and looks more like her father. "Sometimes people don't believe us when we say we are twins," an eleven-year-old Marcia said in an interview. "They think we're just telling a lie."[2] When Millie and Marcia started high school, Amanda had to notify the teachers that the pair were, in fact, sisters.

Marcia and Millie Biggs

In 2018, Millie and Marcia graced the cover of *National Geographic* alongside the tagline "these twin sisters make us rethink everything we know about race." The magazine's special issue explored "how race defines, separates, and unites us," citing how Amanda calls Millie and Marcia a "one-in-a-million miracle" because, although they have the same biological parents, the sisters are often seen by others as members of different racial groups. The magazine asked Amanda what she says when she is asked why this is so, and her simple, go-to answer is: "It's genes."[3]

In one sense, Amanda's answer is spot-on. Though race is a social construct (discussed further in chapter 3), a person's DNA influences a variety of aspects of their appearance. Some physical features, like skin color and hair texture, are racialized—that is, they carry social meaning that can tie a person to a particular racial classification. Marcia's light eyes and sandy hair lead people to see her as White, while Millie's darker complexion suggests a different racial category. The twins inherited separate DNA from their parents to cause these distinctions in their appearances, which, in turn, alter how others racially perceive them.

There is another sense, however, in which Amanda's answer fails: oftentimes the phrase "it's genes" implies "it's *only* genes." Emphasizing the importance of DNA *and DNA alone* is misleading and serves to reify the Destiny Myth, or the flawed idea that the effects of DNA are immutable and inevitable (experts typically call this genetic determinism).[4] So, what makes the Destiny Myth—an idea that has featured prominently in arguments used to justify sterilization and genocide (see chapter 4)—a myth? And when researchers talk about "genetic effects" (or the "effects of DNA") what exactly do they mean?

————

At the heart of the Destiny Myth is the belief that the effects of DNA can be divorced from the social and physical features of the world. Such an erroneous separation traces back to an infamous early pioneer in the field of human genetics named Francis Galton. In 1822, Galton (ironically, like Millie and Marcia nearly 200 years later) was born in Birmingham, England. The son of Samuel Galton and Frances Darwin, he was half-cousins with Charles Darwin, who famously proposed the evolutionary theory of natural selection. As a member of the British aristocracy, Galton received a top-notch education, displaying diverse academic interests from a young age; eventually, he would contribute to scientific fields ranging from statistics to meteorology to forensics.

In 1869, Galton published a book entitled *Hereditary Genius*, in which he argued that a person's "natural abilities" and "mental powers" are biologically inherited. (The word "gene" would not enter the English lexicon until the early 1900s.[5]) In a chapter entitled "The Classification of Men According to their Natural Gifts," Galton writes:

I have no patience with the hypothesis occasionally expressed, and often implied . . . that babies are born pretty much alike, and that the sole agencies in creating differences between boy

and boy, and man and man, are steady application and moral
effort. It is in the most unqualified manner that I object to pre-
tensions of natural equality.[6]

Galton took inspiration from his half-cousin's recently published book on natural selection. Just as a child of towering parents
inherits their biological propensity to grow tall, argued Galton, so
does a child of brilliant parents inherit their biological propensity
for genius. To make his case, he drew pedigree diagrams of eminent British families—the familiar diagram of lines and names
that connects individuals to their closest relatives—parents to
children, parents to grandparents, and so on. These pedigrees, or
family trees, also displayed information about each person's
professional achievements in domains ranging from law to mathematics to poetry. Galton took these generational patterns as evidence of biological inheritance.

The family tree depicted in figure 2, for instance—adapted
from *Hereditary Genius*—displays men in the family of James
Gregory (women are largely omitted from Galton's pedigrees and,
occasionally, identified merely as "daughter"). Gregory, born in
1635, was a Scottish mathematician and astronomer who is credited with designing the first modern telescope. Himself a distant
descendant of a noted mathematician, Gregory would watch
his descendants thrive in the fields of mathematics, astronomy,
and medicine. He had a son, three nephews, a grandson, a great-
grandson, and two grandnephews who attained various prestigious
professorships at the Universities of Oxford, Edinburgh, Aberdeen, and St. Andrews. Gregory's great-great-grandson, Archibald
Alison—perhaps the black sheep of the family—eschewed science and mathematics and instead became a historian. Galton
viewed the Gregory pedigree and others as evidence that scientific
and mathematical genius are biologically transmitted across generations; he largely ignored the patently unequal class structure of
the Victorian era—where the richest 5% of adults held 85% of the
total English wealth—as a potential alternative explanation.[7]

MEN OF SCIENCE.
Pedigree of the Family of Gregory.

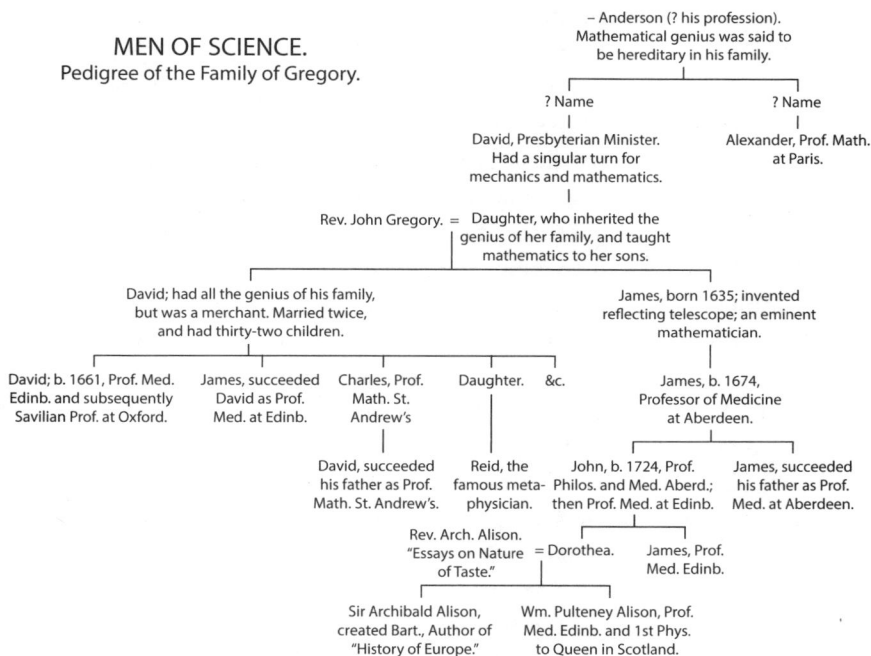

– Anderson (? his profession).
Mathematical genius was said to
be hereditary in his family.

? Name

David, Presbyterian Minister.
Had a singular turn for
mechanics and mathematics.

? Name

Alexander, Prof. Math.
at Paris.

Rev. John Gregory. = Daughter, who inherited the
genius of her family, and taught
mathematics to her sons.

David; had all the genius of his family,
but was a merchant. Married twice,
and had thirty-two children.

James, born 1635; invented
reflecting telescope; an eminent
mathematician.

David; b. 1661, Prof. Med.
Edinb. and subsequently
Savilian Prof. at Oxford.

James, succeeded
David as Prof.
Med. at Edinb.

Charles, Prof.
Math. St.
Andrew's

Daughter.   &c.

James, b. 1674,
Professor of Medicine
at Aberdeen.

David, succeeded
his father as Prof.
Math. St. Andrew's.

Reid, the
famous meta-
physician.

John, b. 1724, Prof.
Philos. and Med. Aberd.;
then Prof. Med. at Edinb.

James, succeeded
his father as Prof.
Med. at Aberdeen.

Rev. Arch. Alison.
"Essays on Nature
of Taste."   = Dorothea.

James, Prof.
Med. Edinb.

Sir Archibald Alison,
created Bart., Author of
"History of Europe."

Wm. Pulteney Alison, Prof.
Med. Edinb. and 1st Phys.
to Queen in Scotland.

FIGURE 2. Pedigree of the Gregory Family. Originally published in *Hereditary Genius* by Francis Galton, Macmillan, 1869.

Galton is responsible for coining the now-familiar phrase "nature versus nurture" to describe genetic and environmental influences, respectively. He did not deny that a person's environment shapes their health, well-being, and life outcomes, but he saw environmental influences as distinct from, and often subordinate to, genetic influences.[8] Galton, and countless others after him, attempted to sever nature from nurture, giving the illusion that DNA somehow operates separately from the social and physical world that humans create. However, genes and environments are not so easily disentangled.[9] The sociologist Dalton Conley poetically illustrates this point by arguing that nature and nurture come together to form a Möbius strip, an intriguing mathematical shape that consists of only a single side.[10]

A Galtonian account of inheritance makes the effects of DNA seem simple, deterministic, and strictly biological—"it's just genes," and those genes operate within the body, whereas social and environmental forces operate outside of it. To Galton, a child seemed consigned to the fate of their biological parents; this belief led him to found the eugenics movement. Derived from the Greek work *eugenes*, which translates to "good in birth," the eugenics movement focused on trying to control human evolution and development using an array of tactics (see chapter 4). Eugenicists like Galton hoped to improve the "biological quality" of humanity. Their ideology would eventually gain widespread attention and adoption throughout Europe and in the United States.

Today, the Destiny Myth continues to both validate existing social inequalities and excuse not helping those in need: consider, for instance, US President Donald Trump. The unpredictable reality-TV star turned politician has long emphasized the deterministic role of DNA. In 1988, he told Oprah Winfrey on her talk show that "you have to be born lucky in the sense that you have to have the right genes."[11] Michael D'Antonio, who penned a biography of Trump, believes that the Trump family subscribes to a so-called racehorse theory of human development,[12] explaining that the Trumps "believe that there are superior people, and that if you put together the genes of a superior woman and a superior man, you get a superior offspring."[13] While running for president and during his first tenure in the White House, numerous video recordings captured Trump making statements like: "you have good genes—a lot of it is about the genes, isn't it" and "some people cannot genetically handle pressure."[14]

———

In this fast-paced age of molecular genetics and polygenic scores, DNA and polygenic prediction are garnering increased power and attention. As they do, the potential impact of the Destiny Myth grows, and the oversimplistic ways that schools teach people to think about DNA become more and more damaging. Unpacking exactly how DNA affects traits can help unravel the dangerous sort of determinism embodied by the Destiny Myth. Modern genomics research deploys a very specific (and perhaps counterintuitive) definition of cause and effect. Understanding this notion of causality—what it means for one factor to affect another—is crucial for making sense of recent genomic discoveries.

Many of us have probably heard the phrase "correlation does not entail causation." Correlation is a concept that describes the statistical relationship between two "things" (i.e., variables), which can be readily observed using data. Two variables are correlated when they tend to move together. Take, for example, (1) the number of glasses of red wine that a person consumes each week and (2) their lifespan. If the results of a survey show that people who drink more wine also tend to live longer, then there is a correlation between red wine consumption and lifespan. (Sam, who does not much like wine, would be unphased by this correlation and will instead stick to drinking beer.)

Causation is a somewhat trickier concept. While it is possible to directly observe correlation, causation cannot be seen firsthand; instead, it is always inferred. Two variables are causally related when one actually *affects* the other. However, it is often difficult to learn whether the relationship between two variables represents causation (rather than mere correlation). A causal relationship between red wine and lifespan would mean that people who drink more red wine also tend to live longer *because* wine-drinking increases lifespan. When a relationship is causal, altering the input (i.e., how much wine a person drinks) induces a change in the output (i.e., a person's lifespan); in this case, changing the amount of wine a person consumes would lengthen or shorten their lifespan accordingly. Add social class into the equation, though, and maybe the survey will show that wealthier people tend to both drink more wine *and* live longer. This case would show correlation but *not* causation, as the saying goes, between wine consumption and lifespan; wine drinkers may simply live longer because they have more money to spend on medical care and other healthy choices. (Daphne, a wine enthusiast, likes to believe that the relationship is, in fact, causal, so her penchant for wine will help her live a long and happy life.)

People make claims about causal relationships every day, even without actually using the words "cause" and "effect." Whenever someone uses language like "leads to," "induces," "makes," "brings about," "prompts," etc., they are making a claim about causality. Though folks frequently make causal claims, they do not always have the necessary data to back them up. To avoid these pitfalls, genomics researchers—and most other scientists, for that matter—utilize a helpful thought experiment to understand causality and interpret the results from their studies.

———

Nearly a decade ago, Sam had a crucial decision to make: should he pack up his bags and move to Palo Alto, California for graduate school, or should he stay in Austin, Texas—the only true home he had ever known—with his closest friends, nuclear family, and girlfriend? Sam ultimately decided to make the pilgrimage across the barren and serene desert of the American Southwest and start work on his PhD, but imagine a parallel universe where he instead chose to stay in Texas. On Sam's most stressful days, he sometimes finds himself romanticizing this alternate version of his life, where he continues waiting tables and aimlessly pedaling around on his single-speed bicycle, trading long days in lecture halls for late nights in concert halls. This daydream, while perhaps unrealistic and overly sentimental, raises a question: What would Sam's life have looked like if he had remained in the Lone Star State? Is there any way to truly know if moving to the West Coast actually increased his happiness?

Daphne often finds herself asking similar sorts of questions. For most of her upbringing, she was an unapologetically unathletic child who detested sports. Instead, inspired by her mother's former acting career in Nigeria, Daphne joined her middle school's drama club. In eighth grade, she traveled to Washington, DC, for a theater competition that sent her life in an unexpected direction: as her troupe drove across the Potomac River, she peered out and saw narrow boats gliding across the water. A member of Daphne's theater troupe, whose older sister was a rower, explained that they were witnessing a sport called "crew." Daphne, enthralled by the graceful movement of the boats, vowed in that moment that she would learn to row. Fortunately, her local public high school had a team, which Daphne joined the following year. Much to her surprise, she proved to be a successful rower, going on to row for Stanford University, the University of Cambridge, and the United States in the Under-23 World Championships. What if, on that fateful day, Daphne had sat on the opposite side of the van, engrossed in a conversation with a friend, and entirely missed seeing the rowers on the Potomac?

Try thinking back to a defining decision in your life. Do you ever wonder what would have happened if you had chosen differently—if you had instead picked the road not taken? What if you had instead applied for that one job over the other? Or, perhaps the key plot twist in your story hinges not on a decision but instead on a chance encounter. What if you had not attended the party where you met the love of your life? The forces that push people's lives toward new, uncharted paths raise plenty of "what if" questions: What if I had not been at that place at that time . . . where would I be today?

To wonder what life would have been like if some key event happened differently is a strikingly human tendency. These "what if" questions are, at their core, questions about how one factor *affects* another—how one choice might cause a particular outcome and another would have led to a different one. Research into the effects of DNA seeks to answer a similar "what if" question: if I had inherited different DNA, how would my life be different?[15] As you will see, the effects of DNA can manifest as the result of markedly nonbiological processes. That is, polygenic scores predict individual outcomes *because of*, rather than despite, social and physical environments. Francis Galton was well off the mark when he pitted "nature" against "nurture." Understanding why "what if" questions are crucial for delineating cause from effect in a clear, rigorous way requires thinking in terms of parallel universes, in which people follow different paths than they do in this reality.

————

Sam and Daphne have a shared love for TV crime dramas—especially those set in the United Kingdom. One of their American favorites is the mid-2000s science fiction series *Fringe*. The show follows a fictional FBI department tasked with investigating mysteries relating to a parallel universe. As the series unfolds and the line between the real world and a parallel one blurs, glimpses of an alternate reality emerge.

Researchers hoping to identify causal relationships between DNA and outcomes use—either implicitly or explicitly—the notion of a hypothetical, alternative state of the world from the one that is immediately observable: a parallel universe, so to speak. This counterfactual way of thinking dates back to eighteenth-century philosopher David Hume and is used by a wide range of scientists to define cause and effect.[16] Conceptually, every event that has ever happened and ever will—each action, decision, or intervention—has a parallel universe where that event *did not* happen and where, instead, an alternate version of you (your "doppelgänger") made a different decision or had a different chance encounter. When using counterfactual thinking, a causal effect is identified by comparing two states of the world: an observable world and a parallel universe.

Did moving to California have a causal effect on Sam's happiness? Did seeing those boats out on the water have a causal effect on Daphne's athletic career? The answers to these questions depend on the outcomes of two different versions of Sam and Daphne: real Sam and real Daphne, the authors of this book, and parallel Sam and parallel Daphne, their respective doppelgängers. Suppose that real Sam, who moved to California, is happier than parallel Sam, who stayed in Texas. Using counterfactual thinking, the additional amount of happiness that real Sam has compared to parallel Sam is considered the *effect* of moving to California on his well-being. Similarly, counterfactual thinking would contrast Daphne's current life to a universe where parallel Daphne instead sat on the opposite side of the van and missed seeing the boats out on the water. If, in this alternate universe, parallel Daphne never picked up an oar, then seeing boats that day *caused* her to become a rower.

There is one key challenge of using counterfactual thinking in practice: comparing differences across *two* worlds. Unlike in *Fringe*, a single world is all that can be observed (i.e., the real world), and everything else is merely guesswork. "What if" questions are all hypothetical; Sam will never know what would have happened in his life if he had decided to stay in Texas, and Daphne will never know what her life would have looked like had she missed seeing those rowers on the Potomac. While it is theoretically straightforward to define a causal effect for any event or decision, actually knowing and proving the magnitude of that effect is not always easy. The parallel universe and the doppelgänger remain objects of science fiction. Or do they?

Statisticians have developed an array of methods and tools that allow for glimpses into parallel universes. When properly applied to the inheritance of DNA, these methods help expose the faults in the Destiny Myth. The most famous of these statistical tools is the randomized experiment. A randomized experiment tests the effect of some intervention or treatment by randomly selecting which individuals receive the intervention (the treatment group) and which do *not* receive the intervention (the control group). The use of randomness means that every person is equally as likely as any other to be assigned to the treatment group and the control group, ensuring that no pattern exists between a participant's chances of receiving the intervention and their characteristics (e.g., their income, gender identity, or personality traits). Because no systematic differences exist between the two groups, the control group can be used to approximate the unobserved counterfactual of the treatment group. In other words, researchers create a world where someone receives the intervention and a parallel world where that person, or someone like them (their doppelgänger, you might say), does not. These two parallel universes allow for an *estimation* of the average causal effects of an intervention, thereby gaining information on causal processes. Randomized experiments are, for example, how vaccines for the COVID-19 virus were validated before being scaled up for wider distribution.

In the case of DNA, though, randomized experiments typically are neither ethical nor possible. Fortunately, random variation allows for "natural" experiments in some cases: a child inherits a random half of their parents' DNA via a process known as genetic recombination.[17] Thus, siblings do not resemble each other to the extent that identical twins do; unlike identical twins, siblings inherit a different random half of their parents' DNA. In fact, this randomness plays a key role in human evolution via natural selection. Comparing DNA differences between sibling pairs, which are necessarily the result of recombination, allows researchers to better identify the causal effects of DNA. While efforts to develop and apply these techniques remains ongoing, such family-based studies are likely to play a key role in the future of genomic research regarding complex traits.[18]

A crucial aspect of counterfactual thinking is that the effect of a cause is defined by comparing *exactly two* parallel universes. These two universes differ only with respect to the causal factor in question, and all other aspects of the world are "held equal"—they remain identical to each other. This fact has important implications for understanding causal effects in general and the effects of DNA in particular.

Suppose, for instance, that researchers used a randomized experiment to estimate the effect of participating in a Head Start–style program on a child's likelihood of graduating from high school. (Head Start is public pre-K program that arose as one of the anti-poverty initiatives that comprised President Lyndon B. Johnson's Great Society.)[19] Imagine that identical randomized experiments are conducted in different Great Lakes cities: Detroit and Toronto. In both cities, children are randomly assigned to either a treatment group, where they attend Head Start, or a control group, where they do not receive any intervention at all. Many years later, researchers follow up with the now-adult study participants. Within each city, they calculate the rates at which members of the treatment group and control group respectively graduated from high school. They take the difference between the graduation rates in the Detroit treatment group and the Detroit control group to estimate the effect of Head Start in Detroit. They do the same with the graduation rates in the Toronto treatment group and the Toronto control group to obtain estimates of the effect of Head Start in Toronto. Interestingly, although attending Head Start was similarly effective at improving the health and well-being of children at age five in both cities, its effects on high school graduation rates were greater in Detroit than in Toronto. Why might this be?

One solution to this hypothetical puzzle deals with how counterfactual thinking "holds equal" everything besides Head Start participation. Because these cities are different (e.g., in different countries), what exactly is held equal in each also differs. Suppose that Head Start improved child development in the treatment group (relative to the control group) in both cities. However, in Detroit, it's possible that students are "tracked" into different level classes based on their academic achievement when they enter school. Toronto, on the other hand, might assign its students to classes irrespective of their initial academic performance. In both cities, children who attended Head Start entered school with improved academic achievement compared to children who did not. In Detroit, though, these initial treatment effects compounded and grew as Head Start students were more likely to be tracked into classes that provided them better academic experiences. In Toronto, which did not track students, the initial treatment effects did not affect the classrooms in which Head Start students landed.

Both the Detroit and Toronto studies identify valid causal effects of Head Start. However, they do so by identifying the causal effects of the intervention *given whether and how students are tracked*. In other words, the Detroit study identifies the effects of Head Start if a child lives in a district that tracks, whereas the Toronto study identifies the effects of Head Start if a child lives in a district that does not. School systems that track students serve to magnify any academic differences between students that exist when children enter kindergarten, which in turn increases the effects of any early childhood intervention (like Head Start) that alters a child's academic achievement. Part of the impact of Head Start in Detroit occurred because tracking meant that the initial academic achievement of a child shaped the educational opportunities they were given later in life.

If Detroit were to start randomly assigning students to classes (rather than tracking them based on their initial academic ability), the causal effects of Head Start in Detroit would decrease to the levels of the effects of Head Start in Toronto. The assumption of holding all else equal means that causal effects are *contextual*, so they apply only to a particular place and time. Most importantly, causal effects identified using counterfactual thinking, genetic or otherwise, are not immutable. These causal effects can only be identified when holding a set of conditions fixed; by changing these conditions, the causal effects may also change. Understanding this simple truth can completely alter a person's outlook on the role that genes play in the world.

Wielding this new conceptual weapon, the counterfactual, provides greater clarity when thinking about the effects of DNA. Over the last decade, the field of human genomics has assembled large genomic datasets and, for the first time, identified relationships between specific DNA variants and a wide range of social, behavioral, and health outcomes. These genetic variants can be condensed into a polygenic score, which aims to summarize how a person's DNA is expected to shape a given trait. But what exactly does it mean for DNA to affect a person's life? The answer is somewhat unexpected.

Using counterfactual thinking, the causal effect of a given DNA variant on any outcome stems from a comparison of the world where a person inherits that variant and the parallel universe where a doppelgänger does not. This parallel suggests that the ways in which social and environmental forces operate on individuals' DNA differences are still technically a causal effect of DNA—but, crucially, an "environmentally mediated" effect of DNA. This thought experiment highlights the fallacy of saying that DNA affects social, behavioral, and health outcomes *on its own*. Instead, social and environmental conditions facilitate the effects of DNA. This relationship is analogous to how, in the Head Start example, a portion of the program's effects in Detroit are mediated by, or operate through, the practice of tracking students based on their initial academic achievement.

Consider the example of sickle cell anemia, a debilitating disorder caused by a DNA variant on chromosome 11. Sickle cell anemia affects the shape and flexibility of red blood cells, which carry oxygen. As a result, the unusually shaped sickle cells get stuck in blood vessels, restricting blood flow and causing severe pain. While this disease may seem like a simple example of genetic cause and effect, in reality, it is more complex. How long and how well someone lives with sickle cell anemia is impacted by the quality of healthcare they receive, including whether they have access to medications to help manage pain or receive medical treatments for the side effects of the disorder (like frequent infections). Therefore, although sickle cell anemia is ultimately "caused" by a DNA variant, factors like access to healthcare mediate the effect of that DNA variant on a person's health and lifespan. Similarly, with modern medical technologies, cystic fibrosis, another harrowing monogenic disorder, no longer imposes the death sentence it once did.

Millie and Marcia Biggs also illustrate the influence of environment on the effects of DNA. Because of randomness during recombination, Marcia inherited DNA that produces lighter skin, and Millie inherited DNA that produces darker skin. DNA largely influences skin color through altering the quantity and quality of melanin pigments in a person's skin tissue. The difference in skin color between Millie and Marcia, while biologically innocuous, may have a profound impact on the trajectories of each child. Imagine that, as a result of colorism, Millie experiences discrimination and racial animus to a greater extent than Marcia. Applying counterfactual thinking, colorism operates on DNA-influenced variation in skin color and creates meaningful differences in, say, mental health between the two siblings. Thus, the colorism pathway is considered part of the causal effects of DNA on mental health.

The existence of colorism in the world is a factor that is held equal in parallel universes when the effect of DNA is calculated. Colorism is technically counted as an effect of DNA because an individual's DNA plays *a role* in the causal process, but colorism and racial discrimination are a central part of *why* DNA variants related to skin tone impact other aspects of a person's life. In a world without racism, the DNA variants associated with skin pigmentation would not, in turn, have the social power to cause differences in life outcomes between individuals. The role of environment is clear. It is unsatisfying, at best, to label racial discrimination a mediator for the effects of DNA. What is lost from much of the communication is that, using counterfactual thinking, an effect of DNA is simply a causal pathway that involves any DNA *at all*. Sure, the DNA a person inherits can influence that person's life; if Millie and Marcia were identical rather than fraternal twins, they would have inherited the same DNA from their parents and would likely appear to be the same race—and *National Geographic* never would have written a story about them. But understanding counterfactual thinking makes it clear that, even when a person's DNA matters for their life outcomes, it isn't destiny.

———

DNA does not operate in a vacuum, but the Destiny Myth continues to feature in arguments regarding social inequality and social policy.[20] Charles Murray, a researcher at a conservative think tank called the American Enterprise Institute, is no stranger to the myth that the effects of DNA are immutable and inevitable. Best known for co-writing the incendiary 1996 book *The Bell Curve*, he has used DNA time and time again to argue that the inequalities of the modern day are biologically inevitable (and largely impermeable to societal intervention). In his 2020 book entitled *Human Diversity*, Murray writes: "It is not within our power to do much to change personalities or abilities or social behaviors by design on a large scale."[21] He eventually makes his position clearer: these constraints are "imposed by human nature." He calls on readers to accept this reality, which should, in his opinion, guide "the operations of just about every social, cultural, economic, and political institution." Murray, who has spent most of his career trying to reduce and eliminate America's social welfare system, wields the Destiny Myth to distract from structural causes of inequality and instead places blame on individuals,[22] and more specifically, their DNA.

Some folks, like Murray, have misinterpreted recent genomic discoveries to purportedly provide evidence for the Destiny Myth (chapter 6 details this topic further). He has argued, for instance, that polygenic scores for educational attainment capture "what is innate."[23] However, the suggestion that polygenic scores capture innate characteristics reveals an unimaginative and utterly incorrect conception of how DNA shapes life outcomes—one that reeks of the Galtonian idea that DNA influences people's personalities, abilities, and health in simple, biological terms that exist separately from the social and environmental aspects of the world. Polygenic scores capture countless different processes, most of which researchers barely understand at this point. Using counter-factual thinking makes clear that polygenic scores are not as straightforward as Murray would like you to think. Polygenic scores do not show "what is innate"; instead, they serve as reflections of the social world.

For instance, DNA can influence aspects of athleticism (e.g., height, wingspan, and propensity for injury). These traits can, in turn, influence the chances of receiving a college athletic scholarship (like Daphne did). DNA can also influence someone's susceptibility to alcoholism, a key risk factor for dropping out of college without a degree. But the exorbitant costs of higher education in America and the ubiquitous binge drinking culture on many college campuses are neither immutable nor inevitable. In a society where college is readily affordable, DNA related to athleticism may not influence whether a person gets to pursue higher education. In a campus climate where responsible drinking is the norm, DNA related to alcohol addiction would be less likely to hamper graduation prospects.

The way DNA operates *through* environmental pathways is not just a feature of socially constructed traits like educational attainment. It is true that academic degrees do not exist in the same way that a physical blockage in the coronary artery does. For more clinical and biological traits, like heart disease, it may well be that certain DNA variants operate through pathways that are largely within the body—for instance, DNA that influences the size and shape of the coronary artery or that regulates cholesterol levels in the bloodstream. However, DNA will still influence a person's heart disease risk via social pathways.

A genomic study on heart disease may pick up DNA variants associated with cholesterol levels, but it will also pick up DNA variants that are associated with educational attainment—because in the United States today, a person's level of schooling shapes their access to healthcare and preventative medicine. Similarly, genomic studies on heart disease will also capture DNA variants that are associated with how stressful people's jobs are, whether they smoke cigarettes, or the frequency with which they exercise. The social context becomes "baked in." A central challenge of the genomic era is that there does not currently exist a rigorous method for identifying through which of the countless potential processes a given DNA variant operates. For this reason, existing polygenic scores for heart disease will capture all processes involving DNA, including the ones that are deeply contextual and environmental (just like polygenic scores for educational attainment).

———

Utilizing counterfactual thinking to understand why the effects of DNA cannot be cleanly separated from the environment is crucial for disrupting the faulty and damaging reasoning underlying the Destiny Myth. The effects of DNA operate by means of the social and physical world, and phrases like "genetic effects" or "effects of DNA" can obscure the fact that there are often many environmental links that exist in the causal chain.[24] Even if someone tries to convince you otherwise, there exists a complex relationship between genetic influences and the social and physical world; polygenic scores will always only offer information that is contextual—about the predicted relationship between DNA and some outcome *in a particular place and time.*

DNA is not destiny. It is not a fortune teller that prescribes a person's fate. Nevertheless, the Destiny Myth continues to proliferate. As the upcoming chapters explain, genetic myths also have real-world repercussions. As genomic technologies advance and DNA-based tools like polygenic scores become more accessible to the everyday consumer, understanding the history and the misconceptions that have produced and maintain the Destiny Myth becomes exponentially more important.

3

The Race Myth

Mildred Delores Jeter was shy and soft-spoken. A tall, slender child, her friends and family affectionately nicknamed her "String Bean." At age eleven, through her older brothers, she first met Richard Perry Loving. Richard, a local drag-racing day laborer six years her senior, initially came across as arrogant to Mildred, but a friendship between the two formed with time, and as these things sometimes go, eventually that friendship evolved into a romance. Richard and Mildred (or "Bean," as he affectionately called her) married in Washington, DC, during the summer of 1958 and soon after returned to their hometown of Central Point, Virginia (not far from where Daphne grew up).

A month after their wedding, the Lovings were startled awake in the middle of the night by police officers bursting into their bedroom. The officers arrested the couple and took them to a jail in nearby Bowling Green; their marriage was declared illegal. A county judge sentenced each of them to one year in prison. Rather than spend time incarcerated, Richard and Mildred agreed to pay court fees (about $350 each in current dollars), leave Virginia, and not return for 25 years.

The problem with the Lovings' marriage was that they were an interracial couple: Richard was White, and Mildred was not. Mildred's father was a mix of Cherokee and Black, and her mother was part Rappahannock. Mildred's own perception of her racial identity varied over her lifetime; however, others typically viewed her

Richard and Mildred Loving

as Black. As a child, the 1940 Census listed her and her family as "Negro."[1] So, when Mildred and Richard were arrested in Central Point, it was because Virginia law enforcement had determined that a White man (Richard) was illegally married to a Black woman (Mildred).

Virginia first outlawed interracial marriage, referred to as the "abominable mixture," in 1691.[2] The original law required that the White partner leave Virginia within three months of marrying someone of another race.[3] Over time, revisions to the law enacted increasingly harsh penalties. When Richard and Mildred married over 200 years later, the Racial Integrity Act of 1924, one of Virginia's strictest Jim Crow segregation laws, had taken effect. Violators incurred steep fines, and both members of the couple could be subject to jail time. Washington, DC, viewed the Lovings' marriage as perfectly legal, but the moment they crossed back into Virginia, they were breaking the law. Forty-one states in the nation at one point enacted anti-miscegenation laws that banned interracial marriage. These laws prevented interracial couples from accessing the same rights as their same-race-couple counterparts.

After their arrest, the Lovings traveled to Washington, DC, and moved in with Mildred's cousins in the predominantly Black neighborhood of Trinidad. They brought with them two sons: Donald, just a few months old, and his older half-brother Sidney (Mildred's son). Their daughter, Peggy, was born soon after, and the family tried to settle into their new reality. However, city living proved an unwelcome change from their pastoral existence in Central Point. Mildred was the daughter of sharecroppers and used to being outside, unencumbered by narrow streets and traffic. In DC, the Lovings struggled to find stable work and housing. They missed home.

In the early 1960s, when Donald was hit by a car and sustained injuries, Mildred and Richard decided they had finally had enough. DC was—in their minds—no place to raise a family. However, to return home to Virginia without fear of arrest, the Lovings would need to fight to have their marriage legally recognized. In 1963, Mildred reached out to then–US attorney general Robert Kennedy, who put her in touch with the American Civil Liberties Union. Thus, began years of legal proceedings that would eventually culminate on the floor of the US Supreme Court. Richard Loving proposed a beautifully simple legal argument for the SCOTUS proceedings, asking his lawyer to "tell the Court I love my wife and it is just unfair that I can't live with her in Virginia."[4]

In 1967, the Supreme Court ruled unanimously in the Lovings' favor, striking down the sixteen remaining state laws that criminalized interracial marriage. (While thirty-eight states at one point had anti-miscegenation laws, widespread support for eugenic policies cooled after the atrocities committed during World War II.) *Loving v. Virginia* paved the way for interracial couples like Daphne and her husband Ali, as well as Daphne's parents, Alex and Toyin, who today live legally married in Virginia—just a few counties away from where Mildred and Richard were arrested. After the SCOTUS decision, almost a decade from when they were first forced to leave, the Lovings finally returned to Central Point. Richard built a house for Mildred and their three children, and the couple lived there for the rest of their lives.[5] Today, a plaque commemorating the Lovings stands outside the Bowling Green courthouse where they were first charged.

Legal and social resistance to marriages like Mildred and Richard's stem from the *Race Myth*: the false belief that DNA differences divide humans into discrete and biologically distinct racial groups. The Race Myth harmfully claims that race is biologically— rather than socially and politically—constructed. It has a hand in how people conceptualize race as well as how we make policies on everything from interracial marriage to public education. Leon M. Bazile, the Central Point judge who first convicted the Lovings in 1958, issued his final verdict with the following explanation:

> Almighty God created the races white, black, yellow, malay, and red, and he placed them on separate continents. And, but for the interference with his arrangement, there would be no cause for such marriages. The fact that he separated the races shows that he did not intend for the races to mix.[6]

This type of belief in the existence of innate and discrete racial differences all too often accompanies a further idea that such differences render one racial group biologically superior to another. Virginia's Racial Integrity Act of 1924 was, at least in part, established because of this pervasive and troubling ideological system. During the *Loving v. Virginia* Supreme Court proceedings, Justice Hugo Black questioned Virginia's legal representative Robert McIlwaine III:[7]

> H. BLACK: *Is not the basic premise [of these anti-miscegenation laws that] White people are superior to the colored people and that they should not therefore be permitted to marry because it might pollute the White race?*
>
> R. MCILWAINE III: *Your Honor, I think that there is . . . I think historically that the legislatures that enacted them had that thought in mind.*

McIlwaine's reluctant admission to the racist motivation underlying Virginia's Racial Integrity Act of 1924 contributed significantly to the Supreme Court's unanimous decision to strike down anti-miscegenation laws. Justice Black and the rest of the Court concluded that these laws did not view members of different racial groups as fundamentally equal to one another. The laws, as the Lovings' attorney Philip Hirschkop explained, "were incepted to keep the slaves in their place, were prolonged to keep the slaves in their place," and still viewed "the Negro race as a slave race."[8] Thus, anti-miscegenation laws violated the equal protection and due process clauses of the US Constitution's Fourteenth Amendment. More than half a century later, *Loving v. Virginia* still stands, and many view a person's right to marry a spouse of another race as settled law. However, legal interpretations are never truly settled. In March 2022, then-Indiana senator Mike Braun suggested that the Supreme Court should reverse key past decisions, like *Roe v. Wade* and *Loving v. Virginia*, which federalize issues that he believes should be left up to the states;[9] just three months later, *Roe v. Wade* was overturned, and women no longer held a constitutional right to have an abortion.

The very same biological conceptualizations of race that motivated historical anti-miscegenation laws exist in the present day. The advent and proliferation of molecular genetic data are leaving some hopeful and others concerned about the impact of the DNA revolution on how people think about race. Could the DNA revolution offer a new opportunity for society to illustrate the complexities of human variation and reject the idea that racial categories are biological in nature? Or will new applications of genetic data, like the use of geographic ancestry categories in research and direct-to-consumer genetic testing services, reinforce crude racial categorization systems and breathe new life into biological understandings of race?

Companies like Ancestry.com, MyHeritage, and 23andMe (which declared bankruptcy in March 2025 but, as of the writing of this book, continues to operate)[10] are offering to sell individuals information about their ancestry. Studies have shown that people who take so-called genetic ancestry tests use them to racially identify themselves and classify others.[11] For instance, White supremacists have used genetic ancestry tests to "prove" their racial purity.[12] Parents have used genetic ancestry tests to justify selecting a racial minority status that they believe will increase their child's college acceptance odds.[13] In short, race and ancestry have a fraught and intertwined history. They are often conflated, making them difficult to disentangle. Troublingly, their conflation bolsters the Race Myth.

To complicate matters further, no universally-agreed-upon definitions exist for race and ancestry. Put ten random people in a room together and odds are that they will offer multiple different answers on how exactly to distinguish between race and ancestry. Even medical experts appear to disagree about the terms; when surveyed, US clinical geneticists held inconsistent understandings of race and ancestry and lacked consensus on how each should be used and for what purposes.[14] Leading researchers in science, engineering, and medicine have repeatedly convened committees to examine and debate the use of race and ancestry in scientific studies.[15]

Race and ancestry may seem at first glance like hopelessly intertwined concepts. This chapter aims to disentangle the distinct meanings of both terms and, in doing so, help you to understand what genetic ancestry tests do (and do not) tell you. The *Spy Kids* movies are a perfect starting point.

———

Sam grew up loving *Spy Kids*. Among the all-time best cheesy children's movie franchises, the films follow the adventures of two young siblings, Carmen and Juni Cortez, agents in the Organization of Super Spies. In the first movie, Carmen and Juni must rescue their kidnapped parents from an evil businessman seeking world domination. One of the series' most memorable characters is Uncle Machete—a brilliant if cantankerous inventor who outfits his niece and nephew with spy gadgets to aid them in their missions.

Machete is played by Danny Trejo, a B-list celebrity who has starred in more than 200 films ranging from *Con Air* to the *SpongeBob SquarePants* movie. *Saturday Night Live* even made a music video about him, with Pete Davidson rapping the chorus—"Danny Trejo! I'm in *everything*." Danny may be the most iconic celebrity many people (including Daphne) wouldn't know by name but have undoubtedly seen on the big screen. He is a tan, Latino man with a booming voice and an infectious laugh. Tattoos cover his muscular arms and chest, and he is perhaps best-known for his signature horseshoe mustache.[16] (Go ahead—take a break from reading and Google him.) Given their shared last name, Sam sometimes gets asked if he is related to Danny Trejo. When he was first posed the question, he actually wasn't sure what the right answer was. Sam has never met Danny, but there are tens of thousands of Trejos scattered across the United States. Sam's family and Danny Trejo both hail from the greater Los Angeles area, though. Since Sam moved away from California as a preschooler, he has been largely absent from the annual Trejo family picnic (a fact that some of his uncles will not let him forget). There are countless Trejos at every reunion—could *the* Danny Trejo have been in attendance all along, unbeknownst to Sam?

Fortunately, Sam's paternal grandfather, Al Trejo, had a passion for genealogy. He kept records about the family's history and even created a detailed family tree. The first Trejo ancestor present on the family tree is Sam's great grandfather, Ezequiel Trejo, born in Zacatecas, Mexico, in 1898. Margarita Santa Anna, Ezequiel's future wife, was born in Chihuahua, Mexico, 14 years later. Ezequiel and Margarita each immigrated to California to escape the chaos and violence of the Mexican revolution. There, they met, wed, and raised two daughters and five sons (including Sam's grandfather). One afternoon, Sam was poring over all the relatives of the Trejo pedigree, from great aunts and uncles to second cousins once removed—literally hundreds of Trejos—searching for

Danny Trejo

Danny Trejo. He found an *Aidan* Trejo, a *Danny* Gutierrez, and even a *Danielle* Trejo! But, alas, no Danny Trejo. Much to his disappointment, Sam was forced to conclude that he was not related to the most prominent Trejo of them all.

Given Sam's dive into his grandpa's genealogical archives, most people would sensibly agree that he is not related to Danny Trejo. A geneticist, however, might beg to differ. While it is typical to think of biological families as distinct units of which a person either is or is not a member, research in human genetics pushes people to conceive of relatedness on a much grander scale. Geneticists like to point out that, if a search goes back far enough, everyone shares ancestors with everyone else. Rather than simply asking *if* Sam is related to Danny, the better question is *how* closely are Sam and Danny related?

The language we often use to describe kinship encourages us to skip over the shared history of humanity—that all humans are related to one another—and instead focus on the differences among people. (After all, who likes to think about the fact that you share the same family tree as your spouse?) For instance, calling pedigree diagrams, like the one Sam's grandfather drew or that Francis Galton crafted, a person's family tree is a misnomer; it disguises the fact that, in reality, all humanity shares just one ancestral tree—*the* Family Tree, which contains all 100 billion humans that have ever lived. It is more accurate to think of Sam's grandpa's short pedigree diagram as a mere branch—a tiny portion of the big Family Tree.

Daphne's family branch extends up past her parents (Alex and Toyin) and reaches her four grandparents (Victoria, Maxim, Mary, and Joseph). This tiny branch then connects to a larger branch of the Family Tree that contains her eight great grandparents, her sixteen great-great grandparents, her thirty-two great-great-great grandparents, and so on. Eventually, the branch will reach the trunk of *the* Family tree—containing, for instance, Daphne's countless ancestors who lived during the agricultural revolution, roughly twelve thousand years ago. Even further down the trunk, the very earliest humans form the roots of Daphne's (and, of course, everyone's) tree. If an omniscient being kept a ledger of every birth since the dawn of *Homo sapiens*, a person could hypothetically collect the tiny family branches of everyone that has ever lived and treat them like an enormous jigsaw puzzle.[17] After finding which branches connect, one could weave them all together—parent to child, cousin to cousin—and recreate the single, inconceivably complex Family Tree. Daphne's and Sam's recent ancestors may occupy different branches, but they nonetheless both share the same roots.[18]

Because ancestry is an ambiguous term that can refer to a variety of different concepts, it is vital to be clear about the *type* of ancestry: for instance, familial ancestry. In genomics, the concept of familial ancestry describes the relationships defined by the enormous Family Tree of humanity. It encompasses the genealogical and biological link between the people who came before us and those who will come after. It is far richer and more complex than the manifestations of ancestry that many people will encounter when they receive their genetic ancestry test results. A specific person's familial ancestry is simply their location in this branching map of genealogical descent, their place in the ever-growing Family Tree.[19] By taking any pair of humans that have ever existed, a function of the forking limbs that connect their unique positions in the Family Tree shows their relatedness. Is Sam second or third cousins with Danny Trejo? No. Given their shared surname, though, and the geographic proximity of their recent ancestors, Sam probably is more closely related to Danny Trejo than he is to most other Americans.

Familial ancestry is a matter of fact about the world. The question of where exactly a person sits within the Family Tree has a single correct answer (although the answer may be impossible to determine in reality). Even as the beliefs, practices, and institutions of society evolve, a person's ancestry remains fixed. Nonetheless, layers of social, political, and economic meaning complicate familial ancestry. For example, political institutions often use familial ancestry to pass on power (e.g., the British monarchy, sultanate of Oman, and chiefdoms of Sierra Leone). People can also have incorrect beliefs about their familial ancestry or the familial ancestry of others—for example, a child born as a result of a hitherto-undiscovered extramarital affair may have the incorrect belief that her paternal caregiver is also her biological father. (In one US study, one in fifty sibling pairs who self-report being full biological siblings are shown via genetic tests to, in fact, be only half siblings.[20]) While there is a true Family Tree, limited access and partial knowledge of exactly what that tree looks like gives scholars at best a blurry, low-resolution snapshot.

———

Race, unlike familial ancestry, is not a scientifically inevitable feature of the world. It is a social process that humans collectively

Danny Trejo   Sam Trejo

The "Family" Tree of humanity

The Family Tree
Contains all 100 billion humans that have ever lived

take part in "doing" together. Having grown up in various parts of the world with a Black mother and a White father, Daphne has not always known how to "correctly" racially identify. Yet at every turn, the world asks her, like it asks everyone else, to provide her race: on the Census, at the doctor's office, when filling out a job application. Imagine that you are handed a form that asks for your race. How does it allow you to describe yourself? Are you given a list of categories, allowed to select more than one option, or given the opportunity to identify using your own words? As Daphne goes through the process of racially identifying herself, society simultaneously goes through the process of racially classifying her; similar to Mildred Loving, when people notice Daphne's skin color and hair texture, they rarely arrive at any conclusion other than "Black." The United States has long subscribed to the so-called one-drop rule, which asserts that a person is Black if they have at least one Black ancestor. Historically, the one-drop rule was enforced either by using information about who a person's parents or grandparents were (i.e., their familial ancestry) or guessing a person's race based on their physical characteristics.

In September 1998, Daphne enrolled in the first grade. Her new school, Mosby Woods Elementary, was—much to her parents' delight—far closer to their suburban Virginia home than Daphne's kindergarten had been. Named after a Confederate commander who led guerrilla campaigns against Union supply and communications lines throughout northern Virginia during the Civil War, the school was just a ten-minute walk from their house.[21] A five-year-old Daphne attended the enrollment appointment with her mother, Toyin, who speaks English with a thick Nigerian accent, and her grandmother, Mary, who doesn't speak English at all. At Mosby Woods, Daphne almost immediately entered a remedial reading program, much to her parents' confusion. At home, Daphne's progress seemed fine, and she already had three years of Montessori school under her belt. Did Daphne's teachers see something that her parents were missing? Daphne's father, Alex, a child of immigrants from the Soviet Union, is not one to avoid taking the bull by the horns. Being the stubborn (and sometimes excessive) advocate for his children that he is, he marched down to the school and demanded answers. When Alex arrived, the school's administrators were surprised to learn that he was Daphne's biological father!

The school had likely made several assumptions about Daphne's family after meeting her mother and grandmother earlier in the year. For instance, the school administrators may have thought Daphne was being raised by a single Black mother and further subscribed to the stereotype that Black mothers are generally poor, lazy, and uncaring toward their children.[22] Today, when surveyed, nearly a quarter of non-Black Americans report believing that Black Americans "just don't have the motivation or willpower to pull themselves up out of poverty."[23] Second, Toyin's Nigerian accent, as well as the fact that she spoke with her mother in Yoruba, probably led the Mosby Woods staff to assume that Daphne was being raised in a non–English speaking household. These inaccurate perceptions of Daphne's family life, and the accompanying assumptions, ultimately could have led administrators to place Daphne into remedial reading, believing that Daphne would struggle with the traditional language arts material.

Both of Daphne's parents immigrated to the United States, but Alex—unlike Toyin—is White. He has blue eyes and, at that time, had dirty-blonde hair. Importantly, he speaks English with the American accent he developed as a child living in New York City. Soon after her father visited Mosby Woods, Daphne was switched out of her school's remedial reading program.

Daphne's life has featured a number of experiences similar to this one. Like Mildred and Richard's children, Daphne and her three younger sisters—Diana, Allegra, and Alexandra—have often noticed how they are treated differently when they are with their father than with their mother (and even differently *still* when they are with both parents). While living in Ukraine in the early 2000s, the Martschenkos often found themselves an object of spectacle while walking down the street. A decade earlier, in Kyrgyzstan, while Toyin was pregnant with Daphne, a White man on the street punched her in the stomach. Shocked, Toyin later told Daphne that she thought that the assault had occurred because the man couldn't believe that it was biologically possible for a White person and Black person to procreate. As one of few Black people around, she had grown accustomed to piercing stares whenever she stepped outside. That morning had been no different; the man and some friends had laughed at her when she first walked by them. Toyin regularly witnessed Black people getting harassed by passersby—especially when in mixed-race couples. But this was the first time she'd been physically accosted. As the Martschenkos experienced, external presentation drastically alters the way that people treat others.

How a person identifies internally and gets classified externally is not fixed. Like Mildred, Daphne's racial identity has not held constant. That most Americans would classify Daphne as Black helps to explain why she tends to identify as Black more often than biracial, even though she has a White parent, too; experiencing racial classification by others can play an important role in shaping a person's identity.[24] However, the social, economic, and political forces that produce and reinforce "boundaries" between racial groups vary across place and time.[25] While Daphne is typically seen as Black in the United States, in other settings, she has been classified differently. After she graduated high school, her family moved to Jamaica for her father's final posting as a member of the US Foreign Service. There, for the first time in her life, Daphne was racially classified as White—despite the fact that her skin tone matched that of many Black Jamaicans. Daphne and her sisters were US citizens with diplomatic immunity and a White father, which the local Jamaicans viewed as synonymous with racial Whiteness. Daphne was White because she was privileged; she was privileged because she was White.

Many people experience alignment between their own racial identity and society's classification of their race. Clearly, this is not always the case. Some nationally representative surveys of

Americans collect data on race both ways: racial identity (i.e., a survey respondent's self-report) and racial classification (i.e., an interviewer's classification of the respondent). In about 3% of cases, the self- and interviewer-reported measures of race do not line up![26] In rare cases, these studies also follow the same respondents over time; they have found that when two *different* interviewers report the race of the *same* respondent, conflicting classifications are made roughly 6% of the time.[27]

A textbook example of this phenomenon is Harry S. Murphy, who arrived at the University of Mississippi as a student in the fall of 1945. Murphy was assigned to the ROTC program at Ole Miss by a naval commander who had mistakenly identified him as White. Although nervous when he first arrived, Murphy's pale complexion and wavy brown hair allowed him to pass as White. However, when the military ended Murphy's ROTC program, he transferred to Morehouse College, a historically Black college. Ole Miss had no idea that it had admitted a Black student until decades later in 1962, when the institution fought to prevent another Black student, James Meredith, from enrolling. Murphy then triumphantly broke the news: "Ole Miss was fighting a battle they had no idea they'd lost years ago."[28]

By focusing on cases where racial boundary lines blur, like Harry Murphy being classified as White—or, in the twenty-first century, Rachel Dolezal identifying as Black despite having two White parents[29]—the complexity of race becomes clear: it is a socially constructed and dynamic process, not something that sits fixed and unchanged in human DNA, despite the ideas that the Race Myth perpetuates.

This truth becomes more apparent when the boundaries between races shift altogether. Decades before Murphy stepped onto the Ole Miss campus, White plantation owners looking to replace their Black laborers brought Chinese sharecroppers to the Mississippi Delta. These Chinese immigrants served as pawns in a broader reassertion of White dominance over Black Americans during the Reconstruction period. As one planter's wife put it, "give us five million of Chinese laborers in the valley of the Mississippi and we can furnish the world with cotton and teach the negro his proper place."[30]

At first, Chinese immigrants were considered and treated as members of the "colored" community. They picked cotton alongside their Black counterparts, occupied a "status very near that of the Negro," and sometimes even intermarried with Black Mississippians. In the 1940s and 1950s, as the Chinese community moved out of sharecropping and into towns, the racial boundaries of the Mississippi Delta shifted. Although few in number (there was just one Chinese Mississippian for every 500 Black Mississippians living in the Delta), Chinese Mississippians started opening grocery stores and accumulating levels of wealth once denied to them in the cotton fields. Witnessing the growing economic power of the Chinese community, many White individuals began to fear that their existing racial hierarchy might soon be upended. They relied on a sharecropping system designed to keep tenants poor and landowners rich. What was to stop the Chinese residents—who knew of and had themselves been exploited by such a system—from forming an alliance with Black sharecroppers?

Ultimately, a realignment occurred whereby the racial boundary lines around "colored" were socially redrawn to exclude Chinese Mississippians. The Chinese community eventually distanced themselves from the Black community, terminating any preexisting financial ties with Black clients and ostracizing those among them who married Black individuals. They began presenting themselves as allies to the White community and its racial segregationist efforts against the Black community.[31] The White community, eager to remove any threat to their power, gradually came to think of Chinese Mississippians as non-Black. Sociologist James Loewen recounts this history in his book *The Mississippi Chinese*.[32] He interviewed a local White Baptist minister, who said: "You're either a white man or a n*****, here. Now, that's the whole story. When I first came to the Delta, the Chinese were classed as n*****s." Loewen then asked if Chinese Americans in the Delta were now considered to be White, and the minister responded, "That's right." Thus, the racial boundaries of the Mississippi Delta evolved, as the socially constructed categories of race historically have done and will continue to do. This fact is a key step toward disentangling race and ancestry. From the moment of their conception, a person's spot in the Family Tree is written into the human species' ledger in indelible ink. Unlike familial ancestry, race is fluid; a given person's racial identity, or society's racial classification system, can and does vary across place and time. Race is all about social meaning and interpretation.

————

Everyone holds beliefs about what it means to "be" or belong to a certain racial group. Do Millie and Marcia Briggs, the mixed-race fraternal twins who often present as different races, truly differ in their racial identities? How would you racially classify them? Perhaps more importantly, how have understandings of race developed in the first place?

The idea that individuals differ in such a fundamental, systematic way that creates homogenous groupings existed well before the term "race" emerged in the English language.[33] Race is a word commonly used in the United States today, but it describes a process of differentiation that long precedes it. For instance, the Ancient Greeks used the word Βαρβαρος, from which the English word "barbarian" is derived, to describe nonnative speakers and noncitizens. Egyptian art categorized people into four separate groups: "Egyptians," "Asiatics," "Nubians," and "Libyans." Words and languages change over time, as evidenced by many now-defunct US census racial categories (e.g., Mulatto, Quadroon, and Octoroon, which appeared on the 1890 US Census). Still, the practice of ascribing people to groups at times seems inescapable.

Race, however, is more than just a process of differentiation. It is a process inextricably tied to concepts of inferiority and superiority. In 1749, Georges-Louis Leclerc, Comte de Buffon—a French naturalist—wrote *The Varieties of the Human Species*, in which he used the word race to describe different human populations, becoming among the first to do so. Buffon argued that diverse climate environments caused physical and cultural differences between groups of humans. Although he believed that all humans were part of the same species (*Homo sapiens*), Buffon argued that White Europeans represented the "original race" and that other races were degenerations of it. The degeneration of the White race, Buffon believed, was caused by poor climate and diet. During the eighteenth and nineteenth centuries some, like Buffon, subscribed to degeneration theory, as it came to be known. Others, however, argued that racial groups represented separate species altogether (as the man who punched Daphne's mother's pregnant stomach in Kyrgyzstan may have believed). Francis Galton, for example, used biological arguments to support scientific racism—arguing that distinct racial groups had separately evolved and a racial hierarchy existed, with Europeans at the top and Africans and Indigenous Australians at the bottom. (One chapter of his book *Hereditary Genius* was entitled "The Comparative Worth of Different Races.") Either way, these theories agreed that a hierarchy of human races defined the natural order and that White Europeans stood at the top. Race became a political ideology for upholding power imbalances.

What soon developed from the application of race to human populations was a sweeping academic enterprise that has come to be known as "race science." Race science was a pseudoscience committed to upholding the Race Myth and proving that inborn biological differences between groups of humans exist. This "field" was built on assigning behavioral traits to individuals with certain physical features, like skin color and head shape. Historically, race science served as the expert witness in countless justifications for colonial expansion, slavery,[34] anti-miscegenation laws, and immigration restriction. In short, "bigotry claimed science as an ally" by using scientific methods to justify racism.[35]

Eventually, the idea that biological differences produced and caused differences between racial groups was accepted as a matter of fact. For instance, although most remembered for penning the phrase "all men are created equal" in the Declaration of Independence, Thomas Jefferson also argued that there are "physical distinctions [between Black and White people] proving a difference of race."[36] The Founding Father described physiological differences in terms of how much hair Black people versus White people have or how frequently each group sweats. He also outlined differences in intellect and personality, writing: "in memory [enslaved Black people] are equal to the Whites; in reason much inferior . . . in imagination they are dull, tasteless, and anomalous." These differences, Jefferson believed, were "fixed in nature."

Over half a century after the abolition of slavery, race science broadened from establishing spurious innate biological differences between Black and White individuals to asserting specious innate biological differences between European "races." As immigrants from southern and eastern Europe began arriving in the United States in the 1910s and '20s, anti-immigration sentiments strengthened. Searching for justifications to restrict immigration, members of the Protestant elite in America began creating biological distinctions between themselves and, for instance, Roman Catholics—many of whom were Irish and Italian immigrants. In *The Passing of the Great Race*, the American lawyer and anthropologist Madison Grant argued that European populations could be divided into "three distinct subspecies": the Nordic or Baltic subspecies (i.e., England and communities along the shores of the Baltic and North Seas), the Mediterranean or Iberian subspecies (i.e., southern Europe), and the Alpine subspecies (i.e., central France and much of eastern Europe). Grant warned that recent waves of immigration to America were inundated with "the weak, the broken, and the mentally crippled." He issued an impassioned plea to his fellow US-born White Americans, whom he identified as descendants of the Nordic race: "If the Melting Pot is allowed to boil without control . . . American[s] of Colonial descent will become as extinct as the Athenian of the age of Pericles."[37]

Race was conceived to authorize long-standing beliefs about who is entitled to privilege and whose powerlessness is justified. The concept was created as a way for society to confer privilege onto some and harm onto others. As Ta-Nehisi Coates, the author of *Between the World and Me*, explained: "race is the child of racism; not the father."[38]

Biological conceptualizations of race have long been wielded to depict discrimination, oppression, and social inequalities between groups as inevitable phenomena that society can do little about. The view of race as a natural, objective, and stable way of separating people is why Mildred and Richard Loving encountered years of legal battles, ostracization, and retaliation. Their love and subsequent union challenged the idea that marriage and procreation should only occur between individuals of the same race. Biological views of race also help explain why Marcia and Millie Biggs graced the cover of *National Geographic*—they became symbols of a new way to think about race.

Race is not just a static collection of ideas inherited from a previous time. Instead, it is a sleek social process that is agile and adaptive to new goals and social dynamics. As molecular genomic data become increasingly accessible, some wonder whether researchers' use of ancestry could become a Trojan Horse—a vehicle that allows racialized biological groupings, concealed as science, to sneak past society's collective defenses.[39]

Before the genomic era, knowledge of the Family Tree remained limited to the availability of oral histories and genealogical records, such as birth certificates, which necessarily only go so far back and have limited accuracy (again, think of the case of the child born as a result of an affair).[40] Today, the availability of genomic data has transformed the ability to view the big Family Tree of humanity. While genealogical records only provided familial ancestry on the order of a few generations, the DNA revolution and molecular genetic data have now allowed scholars to travel back in time and glimpse the big Family Tree with more clarity. What forces govern each person's unique location within it? Who

mates with whom and how many children they have together de-
termines the next generation of humans. Historically speaking,
throughout the roughly 10,000 generations since modern humans
split off from other archaic humans (e.g., Neanderthals), location
has primarily determined who pairs up with who to reproduce.[41]
The so-called First Law of Geography, that "everything is related to
everything else, but near things are more related than distant
things," rings true.[42] For the bulk of human evolutionary history,
the world was bigger, metaphorically speaking. Horses were
domesticated just 5,000 years ago, and modern forms of transportation, like the steam-powered ship and the airplane, only came
online in the past couple of hundred years. In comparison to the
network of social and physical links that make up today's global-
ized society, humans' distant ancestors were far less connected to
one another, so there was less cultural and genetic mixing.[43] Thus,
the Family Tree of humanity came to grow myriad branches,
which roughly correspond to different geographic regions. Even
in the modern era, people who live more closely to one another
also tend to occupy more similar regions of the Family Tree.[44]

There are some important exceptions. Religion and culture also
have and continue to shape who mates with whom. Such practices
have therefore left an important mark on the Family Tree. For instance, although Ashkenazi Jews, like Sam's maternal ancestors,
lived side-by-side with various non-Jewish groups in Europe for
hundreds of years, intermarriage between the groups was uncom-
mon (though not unheard of).[45] Judaism preferences within-faith
marriages, offers few avenues for conversion, and has been subject
to stigmatization by non-Jewish groups in Europe over the last
1,000 years, which may explain the relative lack of reproduction
with others nearby.[46] While some intermixing occurred, Ashkenazi Jews and their Europeans neighbors never became fully integrated populations. As a result, Ashkenazi Jews tend to occupy
their own branch of the Family Tree. Similarly, as the *Loving v.
Virginia* story highlights, until interracial marriage became legal
and socially acceptable, consensual reproduction of individuals of

different races in the United States was less common (even when they were quite geographically close to one another).

The Family Tree plays a key role in showing how DNA gets distributed across the world, making ancestry crucial to the study of human genomics. Individuals, of course, inherit DNA from their parents, who inherited their DNA from *their* parents, and so on. (Unless noted otherwise, this book uses "parents" to describe biological rather than adoptive parents.) Thus, people closer to one another in the Family Tree have a more recent common ancestor and tend to share more DNA with one another. Importantly, many modern genomic studies utilize the structure of a specific branch of the Family Tree, which can cause resulting DNA-based tools, like polygenic scores, to be more accurate for individuals residing on one branch (e.g., Northern Europeans) than individuals from other branches (e.g., Eastern Europeans).

In the past two decades, rapid advances in human genomics have spurred a great leap forward in scientists' ability to form a picture of the Family Tree. Nonetheless, the image is still blurry and will likely remain so. This low resolution is because the genomic data of people who are alive today provides an incomplete map of their ancestors. Some people never have children, and even those that do have children only pass on half of their DNA to each child (which explains why biological siblings have DNA differences). With every new generation of humankind, information about the past gets lost. The more generations that separate Daphne from one of her ancestors, the greater the chance that she inherited no DNA at all from that person—because a person's genome is only so big. Someone that has millions of ancestors cannot inherit DNA from all of them (especially since DNA tends to be transmitted in large chunks). Genomic data allows for the observation of how similar two people's genomes are to one another; in other words, comparing the number and length of the chunks of DNA that two people might share provides an estimate of their relatedness in the big Family Tree of humanity. Nonetheless, because of information loss, genomic estimates of a person's ancestry are still just that—estimates.

———

Countless areas of life place people into groups. Children get sorted into different schools, districts, and classrooms. Some athletes play on junior varsity teams, whereas others compete on Olympic teams. Young or old, optimist or pessimist, rich or poor. Almost everywhere and at all times, someone or something is being categorized. Grouping can be an attempt to instill order or to signify difference and belonging. Past and present, race has been among the most pervasive and durable grouping methods in American society.

With the help of sociologists Marissa Thompson and AJ Alvero, Sam and Daphne recently asked Black Americans their thoughts about Blackness by conducting a national survey on race and genetic ancestry tests. As a part of the questionnaire, respondents wrote about what being Black in America today means to them. Reflections touched upon a range of ideas, from the legacy of slavery to present-day policing to economic hardship to Black culture to Black pride. However, there were also a few complicated answers: some respondents described how what made themselves and others Black was something about their biology and their DNA—that people had "Blackness" in their very blood.[47]

Given historical eugenic notions of race, as well as the modern rise and popularity of genetic ancestry testing, it is not difficult to imagine how some study participants came to this type of conclusion. Indeed, some researchers think it is no coincidence that geographic ancestry resembles race as a system of categorization, arguing that such categories leverage DNA to scientifically support race as a biological construct.[48] Saying that "a person is Black if they have Black traits in [their] blood"—like one of our survey respondents did—misses the true meaning of race. Your DNA may contain information about your place in the Family Tree, but race and racism are much more than genealogy. Race is a social and political categorization system that is constantly being constructed and reconstructed. Race is not only how you see yourself and how others see you; it is also the process through which society reifies these lines of social division.

Today, ancestry continues to function as yet another way to group people. Undistilled, a person's *familial* ancestry is defined by the complete Family Tree—a complex and unwieldy tangle of branches, the very definition of big data. The Family Tree is not just a list of all 100 billion people that have ever lived; it is also a map that charts the intricate web of relatedness that connects each person to another. For instance, you might ask, who is the most recent common ancestor shared by both Daphne and Sam?

Genetic ancestry testing companies have popularized a different, quite coarse form of ancestry. Spit into a test tube, drop it in the mailbox, and in mere weeks you will receive an individualized estimate of your *geographic* ancestry. Genetic ancestry tests assign different chunks of your DNA to populations around the globe by estimating how similar your genome is to the genomes of various "reference" individuals. For instance, your test results might indicate that your ancestry straddles a couple of categories: 24% East Asian and 76% Southern European. In reality, these tests provide information about the Family Tree: where in the world your, say, fifth cousins tend to live (you have way more fifth cousins than you might think).[49] However, due to the difficulty of comprehending and engaging with the messiness of reality, humans tend to use simplifications—like the creation of discrete categories—to help manage vast and cumbersome sources of data. There is nothing inherently wrong with using rules of thumb; strategies for reducing the complexity of information are often invaluable for making practical decisions in a complex world. In practice, though, a few coarse categories dominate, which serve to obscure the intricacies of the Family Tree and perpetuate the idea that humans can be categorized into a small number of basic "types." Moreover, problems arise when people forget or ignore the simplification process and start treating such distinctions as "real."

Unlike racial categories, everyone's unique place in the Family Tree is *not* socially constructed. Questions regarding familial ancestry theoretically have a single correct answer. Nonetheless, when this information gets distilled into more coarse groups (like geographic regions), social ideologies and meaning influence the definitions and applications of information about ancestry. Although genetic testing companies might argue differently, there is no single "right" answer to a person's geographic ancestry percentages because there is no single way to chop up the big Family Tree of humanity. Any ancestry categorization scheme requires humans to make decisions about the number of groups to create and where exactly to draw the lines among groups. Categorizing ancestry into discrete geographic categories will always entail a simplification that ignores much of the richness of the Family Tree.

To see the countless different categorization schemes that could simplify the Family Tree, consider the question of how far to "zoom in" on any given region. Different ancestry categorization schemes can vary in how broadly or narrowly they choose to define a particular ancestral group. Consider, for instance, Sam's ancestors who lived in present-day Spain approximately 300–400 years ago. The DNA that Sam inherited from these ancestors could be described as any of the following: European, Southern European, Iberian, or Spanish. Sam was struck by how much more precisely 23andMe chooses to display information about his Spanish ancestors compared to his Indigenous American ancestors. (A shortage of Indigenous American genomic data may limit how well genetic ancestry tests can distinguish between various different Indigenous American ancestries—but the cultural beliefs of those creating and buying the tests also play a role in how genomic information is distilled and ultimately presented.)

Understanding the complexity of the Family Tree is essential to understanding why the Race Myth is false. Imagine that the Family Tree resembles a mature, gnarled oak tree, with a handful of thick branches emerging from the trunk and countless smaller branches twisting out. Say that we decide to divide the tree into discrete contiguous groups, clustering nearby branches together. Where are the lines drawn? How many groups should be created? Reducing complex systems to simpler categories is a natural human tendency. Because physical location plays an important role in structuring each person's unique spot in the Family Tree, the most common methods for categorizing and labeling ancestry are (sensibly) geographic in nature. Crucially, there is no single "correct" answer to *how* to slice up the Family Tree. Whenever humans take a messy and continuous natural feature and break it up into distinct units, there are social and political processes at play; these processes influence the choices made and the tactics employed. The Race Myth asks you to ignore these processes and pretend that the racial categories in the US Census are "real."

Humans like to layer meaning *atop* the preexisting natural features of the world, which sometimes makes it challenging to see a sociopolitical construct for what it truly is. Think about, for instance, the way in which the surface of the Earth is currently divided into the various countries of the world—and the countless other ways borders could have hypothetically been drawn. The Rio Grande, also known as the Río Bravo, is a great river that runs down from the Rocky Mountains in Colorado and along the southern edge of Texas before terminating in the Gulf of Mexico. Since 1848, the Rio Grande has served as the natural boundary between Texas and Mexico. The Rio Grande, like the Family Tree, is "pre-social"—a natural feature of the world that existed long before humans had a name for it. However, it would be a mistake to say that the mere fact that the Rio Grande exists is singlehandedly responsible for, or somehow justifies, existing boundaries. Doing so completely ignores the social, political, and economic history of the region, specifically the Mexican–American War and the subsequent Treaty of Hidalgo, which brought in the land that now comprises most of the southwestern United States (stretching from the eastern edge of Texas all the way to California).[50] It would be ludicrous to suggest that the location of water flowing over land offers an adequate explanation of the complex and specific geopolitical and institutional arrangement of the world. It is similarly ridiculous when folks suggest that the structure of the Family Tree provides an adequate explanation of the specific way that humans "do" race today in the United States and around the world.

Although geographic ancestry categories result, in part, from a social process, they can be falsely depicted as a natural, objective, and stable way of grouping people. For instance, at the beginning of his second term, President Donald Trump issued an executive order denouncing the Smithsonian museum for featuring an exhibit that promoted "the view that race is not a biological reality."[51] President Trump's executive order was, perhaps, inspired by the work of Charles Murray, who often tries to challenge the idea that race is socially and politically constructed.[52] In *Human Diversity*, Murray writes: "Human beings can be biologically classified into groups by . . . ancestral population. Like most biological classifications, these groups have fuzzy edges."[53] Murray would like to believe that the fact that the categories used by 23andMe resemble racial categories is proof that the Race Myth is no myth at all. He wants you to believe that race is innate, biological, and inevitable and that there is a single way to categorize familial ancestry that is essentially synonymous with the existing racial classification system.

How does Murray conclude that the Race Myth is true? He first points out that a person's geographic ancestry percentages, derived from their DNA, are highly correlated with their racial identity; that is, a geneticist can produce a good guess of a person's self-reported race using their DNA in a complex statistical model. Murray then concludes: "If race and ethnicity were nothing but social constructs, [such an exercise] would be impossible." His argument, however, is a subtle trick, a seemingly scientific sleight of hand. By focusing on whether geographic ancestry (derived from DNA) is predictive of racial identity, Murray deflects the question that truly matters: whether racial categories are an inevitable result of human genetic differences.

The analogy of drawing lines upon the Earth to make borders between countries highlights the absurdity in Murray's reasoning. A key premise of his argument is that if some characteristic (i.e., a person's racial identity) is merely a social construct, then it should be impossible to predict it only using information about the natural features of the world (i.e., a person's place in the Family Tree). However, a counterexample quickly highlights Murray's logical error; the current boundaries between countries are quite obviously social and political constructs. Granted, it would be possible—even downright easy—for a researcher to accurately guess which country a plot of land rests within using only information about its natural features (for example, its climate, topography, and geochemistry). Murray ignores the fact that many social processes are layered upon the various features that comprise the world. Simply that society, in part, uses the Family Tree to "do" race does not diminish race's status as a socially and politically constructed phenomenon. Racial categories are only "real" in the same way that the United States and Mexico are "real": because humans collectively place meaning and value into them.

Although there is no single way to distill the Family Tree, imprecise and broad geographic categories tend to dominate how people make sense of humanity's complicated history. But the categories used by many researchers and by companies like 23andMe or Ancestry.com obscure the Family Tree's intricacies. These categories are frequently conflated with racial groupings to proliferate the Race Myth and distract from the racism that gave birth to race. This fact makes it critical to ask how and why any specific categorization regime is being utilized in a genomic study, or for any other purpose for that matter. It is important to interrogate whether the formation of specific groups based on DNA furthers sociopolitical agendas, along with how groupings are used to harm some and benefit others.

A complex and thorough understanding of familial ancestry and the big Family Tree of humanity is a vital tool for combatting the misuse and misappropriation of new genomic tools. Familial ancestry may be a natural feature of the human species—a continuous and messy web that leaves an imprint on the human genome. However, the simplification of familial ancestry into discrete groups, like geographic categories, is a social process akin to the construction of racial categories. The consequences of creating such groups cannot slip through the cracks. As illustrated in part 2, these systems of categorization can irrevocably affect people's lives and, in some cases, even become matters of life or death.

PART II
Debates and Disagreements

4

The Effects
of Genetic Myths

SAM: *Daphne, you're the person who first inspired me to really consider the problematic ways we tend to think about genes and DNA. Do you remember what first sparked your interest in genetic myths?*

DAPHNE: *I can't pinpoint a specific moment; it happened gradually. If I had to guess, many years of life experiences sparked my interest in genetic myths. I mean, I have this distinct memory of a doctor telling me that, because I'm Black, it wasn't concerning that my creatinine levels were elevated. I didn't think anything of it until I was heading back home and had a sudden flurry of thoughts: why did the doctor think it was normal for Black people to have higher creatinine levels? Maybe the doctor would've been concerned if they'd known I'm biracial. Were my kidneys actually working okay? These types of questions definitely inspired me to delve further into genetic myths. How have genetic myths appeared in your life?*

SAM: *Genetic myths pop up when I'm talking to people about my research—when people make assumptions about how or why genetics affect certain traits. If I say I'm studying how DNA influences BMI, people often assume these effects operate through physiology and metabolism, overlooking genetic effects on behavioral factors like diet and exercise.*

DAPHNE: *I've seen them in my research too. I've heard teachers describe Black students as less smart and less capable in ways that suggest those differences are inherent. When I was a grad student, I read this paper by the sociologist Ruha Benjamin. It opened with "the stories we tell matter. They produce meaning and material to build (and destroy) what we call the real world."[1] The idea has stuck with me ever since. Genetic myths are stories that matter.*

SAM: *This all reminds me of the unethical experiment by Rosenthal and Jacobson that we described at the start of the book. Teachers' beliefs about their students, even misguided ones, end up affecting what happens in the classroom. So, even though genetic myths are abstract ideas, they become tangible and real through the way they capture our attention and influence our actions.*

DAPHNE: *And there's also no single way to tell or interpret a genetic myth. They adapt to fit the place and time, and different people will interpret them differently depending on the context.*

SAM: *From that perspective, genetic myths have something in common with biological organisms—they must constantly evolve to survive.*

———

Summers in the South are known for their oppressive heat. The air sits thick and heavy with humidity, and mosquitoes quickly multiply. Charlottesville, Virginia—located about 100 miles southwest of Washington, DC—is no stranger to sweltering summer days. On one such Charlottesville afternoon, in July 1906, Emmett (Emma) Adeline Buck went into labor. She gritted her teeth through the pains of childbirth without any family by her side. Though Emma was married to a local tinsmith named Frank, he had taken off not long after their wedding. Emma gave birth to a baby girl and decided to name her Carrie. Soon after Carrie was born, Emma received word that Frank had died.

With little money or social support, Emma struggled for years to make ends meet and care for Carrie. Soon, rumors began to spread that the single mother was an alcoholic and a prostitute. Eventually, the local authorities stepped in. They removed Carrie, now four years old, from Emma's care and placed her into the foster system. A Charlottesville couple, Alice and John Dobbs, took Carrie in and enrolled her in a local grade school.

At about the same time, a new institution opened its doors: the Virginia State Colony for Epileptics and Feebleminded. In 1910, the Colony, located about an hour's drive southwest of Charlottesville in a town called Lynchburg, was the largest institution of its kind.[2] It was housed in a wide two-story red brick building. Many rooms contained a large, mullioned window that overlooked a sprawling lawn. The idyllic beauty of the Colony may have obscured the institution's sinister purpose: to isolate and detain America's "undesirables."

In grade school, Carrie proved to be a middling student, earning a mix of As, Bs, and Cs. After the sixth grade, she was forced to leave her education behind. Carrie became a full-time domestic servant to her foster family, assisting with chores in the Dobbs household instead of continuing on to middle school. One day, when Carrie was sixteen, Alice and John briefly left town to tend to a sick relative. Perhaps sensing an opportunity, the Dobbs's nephew came to the house. Finding the young girl alone, he proclaimed that he would marry Carrie and then raped her. The nephew left as abruptly as he had arrived. Shortly after her seventeenth birthday, Carrie discovered she was pregnant.

Carrie's pregnancy posed a huge problem for her foster parents; the Dobbs's could not have an unmarried pregnant girl in their household. Her growing stomach would soon raise questions, and those questions would threaten their ability to take in other foster children—a key source of their income.[3] Desperate to find a way to cover up the scandal, Alice paid a visit to a local social worker, then to a nurse, then to several doctors, and finally to a municipal judge who—as it turns out—was a family friend. Rather than mention the rape, Alice instead falsely accused Carrie of deliberately getting herself pregnant out of wedlock. Slowly, a potential solution to the Carrie problem emerged: once Carrie gave birth, Alice and John would take in her baby. Then, Carrie would be gone from their lives once and for all.

In March 1924, a seventeen-year-old Carrie gave birth to her daughter, Vivian Elaine Buck. Just a few months later, Carrie was forcibly separated from Vivian. On a hot summer day, not unlike the day she herself was born, Carrie became inmate #1692 at the Colony—now in its second decade of operation. As part of the usual intake process, Carrie was examined to assess her condition. The Colony's top doctor noted that she was White with dark hair. He declared her physically fit, free of syphilis, and able to read, write, and keep herself tidy. Then, the doctor issued his official diagnosis: Carrie was a "moron," one of the three categories of feeblemindedness in use in the early twentieth century (figure 3).

So far, all of the major decisions in Carrie's life had been made by others. The local Charlottesville authorities took her from her mother and placed her into foster care. Alice and John removed her from school and decided that she would instead work full-time in their home. The Dobbs's nephew, through his rape, wrenched away Carrie's bodily autonomy—robbing her of the decision of when and with whom to have a child. After all of this trauma, a physician that Carrie had just met determined that she would be institutionalized indefinitely. Carrie never had control over her life, but the tailspin was just beginning; things would only go from bad to worse.

In 1924, the same year that Carrie was taken to the Colony, Virginia became among the first states to legalize involuntary sterilization of the "unfit." There was ample leeway in deciding who exactly to sterilize. A 1929 Report entitled "The Legal Status of Eugenical Sterilization" outlined ten qualities that made a person eligible for forced sterilization: (1) Feebleminded; (2) Insane; (3) Criminalistic; (4) Epileptic; (5) Inebriate (including drug-habitués); (6) Diseased; (7) Blind; (8) Deaf; (9) Deformed; and

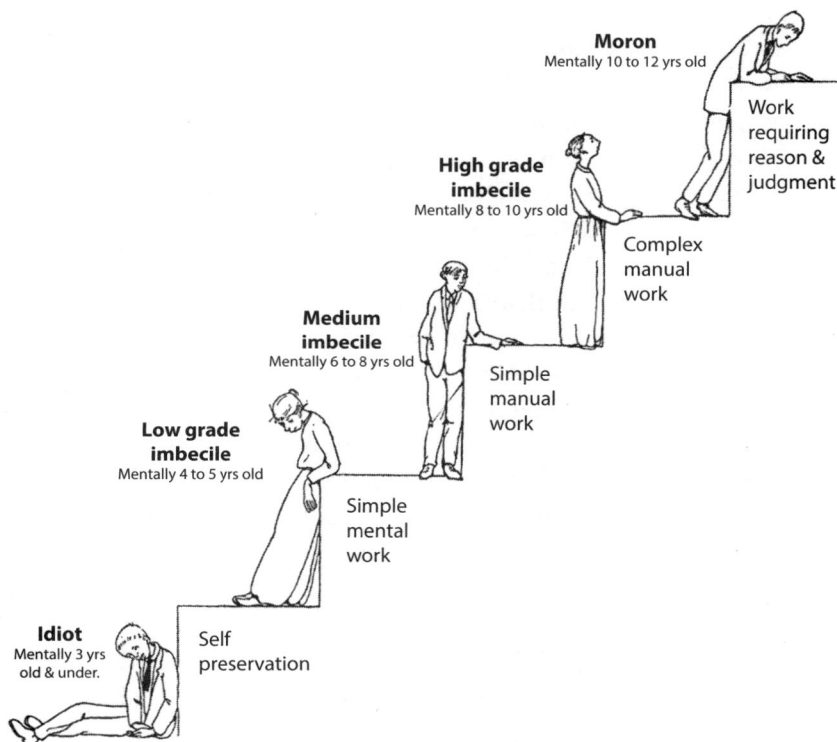

FIGURE 3. Categories of Feeblemindedness. Originally published in *The Survey*, October 11, 1913. Charity Organization Society of the City of New York, 1909–1937.

(10) Dependent. With these criteria, many already-marginalized groups—the poor, communities of color, Catholics, orphans, and the physically and mentally disabled, to name a few—were disproportionately targeted.[4] Proponents of involuntary sterilization feared that the unfit would inevitably pass down their inferior biology, thus perpetuating traits like low intelligence, overactive sex drives, and moral degeneracy in future generations.[5] Thus, they argued, sterilization was a necessity for maintaining the overall welfare of society.

To Colony physicians, Carrie seemed an obvious candidate for sterilization. After all, she was a diagnosed moron, and the fact that she had an illegitimate child signaled she lacked self-control. An even more damning piece of evidence (in the eyes of her doctors) was that, much to Carrie's surprise, her mother, Emma, was also institutionalized at the Colony, having arrived several years earlier. Like Carrie, Emma had given birth out of wedlock and was diagnosed with feeblemindedness. The Colony's leadership took this fact as clear evidence that Emma had biologically passed her feeblemindedness down to Carrie. They believed that Carrie had now passed the condition onto her daughter Vivian. When a social worker assessed just months-old Vivian, she declared that the baby had "a look about it that is not quite normal."[6]

Albert Priddy, the Colony's superintendent, considered it his moral obligation to prevent more social problems like Emma, Carrie, and Vivian from entering the world. He took his job seriously. Believing that Carrie's biological makeup was a threat to society, he filed to have her surgically sterilized under Virginia's new sterilization law (for which he had vociferously campaigned).

Whether the new Virginia sterilization law was technically permissible under the US Constitution was still a legal grey area. Priddy, who considered himself a man of science and reason, did not want his hard work undone by a legal challenge. So, with colleagues, he decided to present what he believed to be a slam dunk case to the courts to solidify Virginia's law: Carrie Buck. Though the legal proceedings were filed in her name, Carrie had no voice in the process. As in countless times during her life, control of her own fate ended up in the hands of more powerful and privileged individuals.

The Colony went from court to court, presenting Carrie's case—serving as both the prosecution and the defense. At each successive judicial level, the courts upheld the sterilization order. Carrie's Colony-appointed legal representatives kept challenging the courts' decisions, thereby pushing the case higher and higher in the court system. Eventually, the US Supreme Court accepted the case for review. A Supreme Court ruling in the Colony's favor

Emma, Carrie, and Vivian Buck

would help ensure that legal challenges to involuntary sterilization from a lower court would not hold water. If the Supreme Court upheld Carrie's sterilization order, involuntary sterilization orders across the country would be legally permissible.

In an eight to one ruling, the 1927 case *Buck v. Bell* found forced sterilization for eugenic purposes to be perfectly legal. In the Court's majority opinion, Justice Oliver Wendell Holmes penned a line that would reverberate for years to come. In reference to Emma, Carrie, and Vivian, the Justice proclaimed: "three generations of imbeciles are enough." With this ruling final, Carrie was sterilized. Priddy, who contracted cancer and passed away before the Supreme Court's ruling, never got to see the culmination of his grand legal scheme.

Following *Buck v. Bell*, the Colony continued to be a prominent site for surgical sterilizations. Over the twentieth century, thirty-three states enacted compulsory sterilization programs, ultimately sterilizing more than 70,000 women without their consent (and many without their knowledge), some of whom are still alive today.[7] One unknowing victim of sterilization was Carrie's half-sister, Doris. For years, Doris struggled to conceive; late in life, she learned that there had never been any hope, as that choice had been taken from her long ago. To this day, *Buck v. Bell* has not been overturned.[8]

In the early days of sterilization, its supporters—predominantly White, upper-class Protestants—fixated on the idea of preventing impoverished Southern and Eastern European immigrants from reproducing. Instead of evaluating the role of phenomena like America's piecemeal social safety net, those with privilege and power may have preferred to believe the Destiny Myth: that those in dire straits had inborn inadequacies that society could do nothing about. In fact, the prevailing narrative was that society only had two options at its disposal. The first was to prevent unfit individuals from passing their deficiencies onto future generations, which prompted sterilizations of women like Carrie. The second, advocated for by the pioneering eugenicist Francis Galton, was to encourage the socially desirable to grow their families. At a fundamental level, these approaches aligned with the primary goals of eugenics: to better humanity and relieve suffering.

So-called Better Baby contests became a key method for encouraging reproduction among America's more desirable members. Nurses in crisp uniforms measured and examined every inch of young infants as a panel of judges awarded trophies to the most "scientific babies"—those who were deemed healthy and strong.[9] Eventually, "Better Baby" contests morphed into "Fitter Family" contests in which entire families were judged using similar criteria. Many have connected such eugenic pageants to modern-day beauty pageants. In fact, as late as the 1930s, the Miss America beauty pageant required contestants to formally provide ancestry information for review, showing that they were members of "the White race" (in addition, those with ancestral connections to the Revolutionary War and *The Mayflower* were given an advantage).[10]

Although initial sterilization efforts in the United States focused on impoverished White Americans, members of other vulnerable communities became targets before long. Countless racial minorities were sterilized, with the Race Myth and the biological threat of non-White reproduction cited as key justifications. Taking inspiration from the United States, Nazi Germany embarked on a genocide against Jewish people, the likes of which had never been seen before. Forced sterilization, along with other eugenic policies, may have lost favor in the aftermath of World War II, but the ideas underlying the eugenics movement didn't abruptly disappear. Scholars have traced the subtle evolution of eugenic thinking, revealing its continued legacy in terms of how people think about health, ability, and of course, DNA.[11] The ideological legacy of twentieth-century eugenics buoys destructive genetic myths that—to this day—many continue to view as ground truth.

Genetic myths are an inherited sort of conceptual acid, and they surface and circulate in connection with noxious claims about human difference. The previous two chapters explained and debunked the Destiny and Race myths, but such myths still remain a durable feature of the social world. Many people believe that genetic myths are fact rather than fiction; faith in these fictions impacts people's lives. Sometimes, these impacts are irreversible.

———

Unlike Virginia's muggy, hazy summers, summers in Scandinavia are bright and temperate. The sun casts seemingly never-ending sunshine over the lush greenery and serene coastline, the best antidote for the lingering effects of a long winter. Still, hate can corrupt even the loveliest of summer days. On the afternoon of July 22, 2011, a twenty-three-year-old Norwegian man rigged his van with a bomb and parked it outside of a government building in downtown Oslo, Norway. Wearing a homemade police uniform, he exited the vehicle, lit the fuse, and walked away. The ensuing explosion left eight dead and more than two hundred injured. Mere hours later, he boarded a ferry to Utøya, a small lake island northwest of Oslo. There, the youth wing of Norway's center-left Labour Party was hosting their annual summer camp. Still impersonating a police officer, he pulled assault weapons out of his bag and began firing indiscriminately. By the time he surrendered to police over an hour later, he had murdered sixty-nine people on the island, many of them children. His rampage has been described as the worst attack on Norwegian soil since World War II.[12]

In the face of such acts of cruelty and malice, it is only natural to wonder about the murderer's motivations. What could lead a person to do such a thing? Just before his spree, the killer emailed a 1,500-page screed—partially plagiarized from the Unabomber's infamous 1995 manifesto—describing his rationale and detailing how others could replicate his horrid crime. He justified his violence as a necessary defense against the terrible threat that Muslims and multiculturalism posed to Norway and Western Europe. Many of his ideas drew on "great replacement theory," a confused and toxic ideology that had arisen in France in recent decades. The French nationalist Renaud Camus coined the term in his 2011 book *Le Grand Remplacement*, describing how increased immigration rates of Muslims from Africa to France were causing a slow "genocide by substitution" of White Christian French culture. For Camus, Islamization and the "genetic manipulations" stemming from immigration were eradicating the essence of French culture and the White race. Such dangerous ideas played an instrumental role in the Norwegian murderer's thinking. At his trial, he delivered a Nazi salute while holding a sign that read: "Stop your genocide!!! Against our white nations!!!"[13]

Eight years later, on March 15, 2019, another man decided to take up the call to arms detailed in the pages of the Norwegian killer's screed. He attacked two mosques in Christchurch, New Zealand, killing fifty-one people. He too issued a hateful proclamation online, directly citing the Norwegian's rant and drawing on great replacement theory.[14] The assault was livestreamed through a helmet-mounted camera.

The video of the shooting went viral, eventually making its way to the computer screen of a teenager in New York. On May 14, 2022, that teenager murdered ten Black Americans in a Tops Supermarket in Buffalo, New York. Like the New Zealand shooter, he livestreamed his attack. Following past examples, he published a 180-page manifesto detailing his motivation and his plans. Scattered throughout its pages were many of the same racist and xenophobic ideas articulated by his predecessors, including, once again, Camus's great replacement theory, which warned of the biological and cultural extinction of the White race.[15] Like the others, the Buffalo shooter saw murder as a necessary act against White genocide. However, he directed most of his wrath toward Black and Jewish people (while the Norwegian and New Zealand shooters had primarily focused on Muslims), and he went to extraordinary lengths to try to assert the biological superiority of White people. The Buffalo shooter argued that races reflect different subspecies of humanity; the Race Myth formed the bedrock of his hateful ideology.

Describing these events is not easy or desirable, but it is necessary to show just how high the stakes are.[16] However baseless, genetic myths can have real repercussions. In a parallel universe without the Destiny Myth, would so many people in America have been forcibly sterilized? If the Race Myth did not exist, would the Buffalo shooter have embarked on his racially motivated killing spree? It may never be possible to answer these questions definitively. In both cases, though, it's clear that genetic myths were at the scene of the crime.

Justice Holmes had the Destiny Myth in the front of his mind when he wrote in his decision that "it is better for all the world, if instead of waiting to execute degenerate offspring for crime . . . society can prevent those who are manifestly unfit from continuing their kind."[17] If Holmes had considered that DNA works through, rather than despite, their environment (as well as the lack of evidence that the traits in question were meaningfully influenced by one's biology), then he may not have supported sterilization. It wasn't Carrie's biological makeup but instead society's lack of commitment to the poor and powerless that threatened to consign her future children to a grim fate.

For the Buffalo shooter, the Race Myth appeared to be a key justification for his crimes. It is, of course, possible that the shooter could have looked for justification elsewhere—that evil will seek out evil. Taking a step back, though, it is easy to see how the Race Myth serves to harden the already toxic great replacement ideology. Without the Race Myth, Camus's great replacement theory largely (and still problematically) takes issue with non-Western cultures, religions, and values—not biology. However, if the Race Myth prevails—if Black and White are not considered made-up categories on a census form but are instead believed to refer to different human subspecies—then, the issue isn't with the beliefs or values that people have but, rather, the DNA in their cells. It was this very strain of thinking that led people to take the extreme measures that resulted in the Holocaust, among other atrocities.

Genetic myths do not just feature in the foundational ideologies of mass shooters and mass atrocities. They also circulate subtly—shaping people's everyday beliefs about themselves and others. Toward the end of Daphne's doctoral fieldwork in Chicago, she visited a social studies classroom in a public charter school and observed a group of fifth graders learning about Katherine Johnson. You may know Johnson's history from the film *Hidden Figures*, which the fifth-grade students had just taken a field trip to see in theaters. *Hidden Figures* tells the story of NASA's "human computers"—mathematicians who calculated the trajectory of spacecraft for key American space missions during the Cold War. Though today, Johnson, played by Taraji P. Henson in the movie,

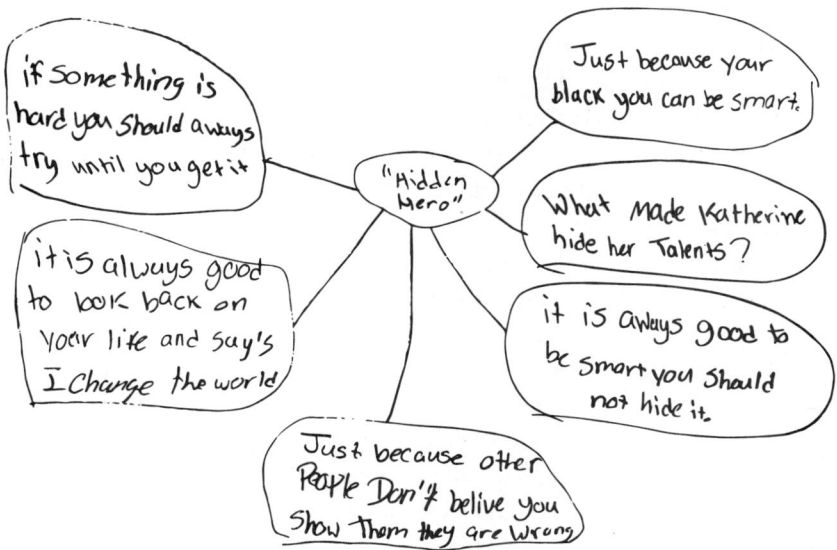

FIGURE 4. Brianna's mind map, April 2017

is recognized as being among the first Black women to work as a NASA scientist, many of her contributions long went unrecognized, her name literally left off of the reports she helped author.

As Daphne joined a small table, students took turns sharing their thoughts about a reading on Johnson. Brianna (named changed for privacy), a young Black girl, shared her diagram of key takeaways. In the top right corner, she had written: "*Just because you are black, you can be smart*" (figure 4). To Brianna, Johnson defied the pervasive and harmful belief that Black people are less intelligent than other racial groups. It is not hard to imagine where she may have internalized such an idea. American teachers systematically expect less of their Black students; indeed, one in four believe genetics can help explain why White students, on average, outperform their Black peers.[18]

As DNA-based tools become increasingly present in day-to-day life, both the impacts of DNA *and* genetic myths must be taken seriously. Both the literal acid in human cells and the conceptual acid of harmful genetic myths affect real people in the real world. Awareness of these dual acids will require simultaneously thinking both big *and* small. Questions about the effects of DNA require people to zoom in, to the very center of cells, and use molecular and statistical methods to interrogate the impact of inheriting one DNA variant rather than another. Researchers like Sam have been working hard to find answers to these kinds of questions. In comparison, questions about the effects of genetic myths require people to zoom out to consider huge, unwieldy questions about our collective history. Researchers like Daphne dedicate their careers to determining the answers to these questions. Both kinds of acid, and the questions each raise, lie at the heart of debates about whether and how to conduct social genomic research and, more broadly, the role that DNA should be allowed to play in our collective future.

———

DAPHNE: *When I hear a student like Brianna say, "Just because you're Black, you can be smart," my heart breaks because she seems to have received the opposite message at some point in her life: that you can't be both smart and Black. Thankfully, one thing she took away from learning about Katherine Johnson's life was that this idea was false.* **Carrie Buck's sterilization and Brianna's mind map demonstrate the persistence of genetic myths and their enduring effects.**

SAM: *I wonder how many girls Brianna's age have internalized similar beliefs about being less capable and less worthy, and have gone uncounted?*

DAPHNE: *I mean, I was one of those girls. I was Brianna. When a teacher says: "I think sometimes people are inherently good at things and some people will never achieve,"*[19] *like I heard during my PhD research, I see the Destiny Myth at work. This is why I am perplexed when colleagues argue that genetic*

*myths are not as pervasive or impactful as I view them to be.
Many folks seem to agree that these myths exist, but they seem
to disagree about whether their existence "matters" in some
tangible way, or what can be done to dispel them.*

SAM: *I completely see where you're coming from, but I also think*
**there's a limit to how much we can learn from individual
stories about the impact and magnitude of genetic myths
on society today.** *What do you think drives disagreement
about the power of genetic myths?*

DAPHNE: *Well, I think a key challenge is how diffuse and
widespread the impacts of genetic myths are. It's really difficult
to quantify the total power they have in our world. Sometimes
genetic myths can be hard to recognize because the precise
way that they are articulated, and the stakes they raise, differ
depending on the context. On the surface, the goal of boosting
students' perceptions of themselves seems quite distinct from
the goal of reducing the number of violent hate crimes. In
reality, genetic myths have their fingerprints on both issues.*

SAM: *I can absolutely see how the wide-reaching impacts of
genetic myths, from mass murder to political rallying to a
young Black girl's sense of self-worth, make tallying the harm
they cause difficult. The task of systematically assembling
evidence on this front offers a key challenge for the next
generation of researchers.*

DAPHNE: *Well, do you think writing this book together has
changed your mind about genetic myths?*

SAM: *In some ways, definitely. I can see more clearly now how
genetic myths are important for understanding our subjective
experiences—how and why we each refract the world the
way we do, including the way we view ourselves and others.
In other ways, I'm not so sure. To what extent do we want to
point the finger at harmful ideologies rather than specific
problematic policies and actors? As a policymaker, what
can practically be done about ideologies that are viewed as
problematic?*

DAPHNE: *Yeah, I can see that there is something unsatisfying about blaming a nebulous set of beliefs for a real harm brought on by real people, like the Buffalo shooter. Genetic myths shouldn't be used to excuse violent behavior.*

SAM: *On top of that, I worry that there is a limit to the effectiveness of targeting belief systems (as opposed to the more proximal sources of harm that they may help fuel). The spread of the Destiny and Race myths feels like a fundamentally different kind of social problem than, say, the number of Americans without access to adequate healthcare. If a set of beliefs has produced harm over and over again throughout history, to what extent, and how, should society intervene?*

DAPHNE: *You're right. It's hard to change belief systems, but I don't think that should stop us from trying. Genetic myths are harmful stories, and I think society has a responsibility to reduce that harm. The goal is not to eliminate storytelling—it's to craft new narratives that can help build a more just and equitable world.*

5

The Effects
of Genomic Research

SAM: *Genomics research, and especially social genomics research, can be controversial. In the case of social genomics, I think the controversy centers on a tough question: what is the overall impact on our world when we study the effects of DNA on social and behavioral traits? This question is so large and unwieldy that it is difficult to answer. It's perhaps not as straightforward as answering the question of whether inheriting a certain DNA variant tends to increase, say, a person's body mass index. When people don't have definitive evidence to go on, they have to make best guesses—which often rely on fundamental or untestable beliefs about how the world works.*

DAPHNE: *Agreed. We've definitely learned that we draw upon different guiding principles when considering the impact of social genomics research now and in the future. When I think about the impact of social genomics research, I look back at history as an indication of what is to come: risk begets more risk, harm begets more harm, and myths about genes beget more myths. In my mind, there would need to be a radical shift in how our society is organized and in our dominant ways of thinking to have social genomic research help dispel, rather than perpetuate, genetic myths. I don't see signs of that happening.*

SAM: *But what can bring about a radical shift that starts to undo our wrongheaded beliefs about DNA? I find myself having faith in the idea that—on balance—science and reason are the best tools we have for combating inaccurate beliefs and misconceptions.*

DAPHNE: *Well, let's take a look at how social genomics research currently intersects with genetic myths—and what the scientific community might do about it.*

――――――

Assessing the potential hazards of genomics research means thinking about not just how it could be misinterpreted but also who exactly is doing the misinterpreting. As Carrie Buck's forced sterilization and Mildred and Richard Loving's arrest highlight, prejudiced actors have a long history of twisting scientific language and findings to sustain genetic myths. Perhaps, then, it is no surprise that modern genomics research is the latest content to be appropriated to this end.

Stefan Molyneux—a fringe political commentator, scientific racist, and self-proclaimed men's rights activist—has had a lot to say over the years. Since 2005, when he first began recording his podcast *Freedomain Radio*, he has released thousands of hours of tape. His broadcast covers a host of social and political issues that range from gender roles (episode 2,264: "How Feminism Destroyed Europe") to free speech (episode 3,239: "Social Justice Warriors Always Lie") to climate change (episode 3,508: "The Global Warming Hoax"). Though YouTube banned his channel in 2020 for violating the company's hate speech policy, Molyneux still releases content (largely guest interviews and monologues) nearly every day through his personal website.

In most respects, the 3,697th episode of *Freedomain Radio*—entitled "Will Genius Be Genetically Engineered?"—is no different from the show's myriad other episodes. In this May 2017 interview, Molyneux muses about one of his all-time favorite topics: genes and average differences in intelligence across racial groups. Molyneux, described in the British newspaper *The Independent* as an "alt-lite philosopher with a perverse fixation on race and IQ,"[1] has said numerous times that he does not "view humanity as a single species."[2] In episode 3,697, he samples from his usual accounts of the ills of society, bemoaning, for instance, the countless social justice warriors who use "crazy things" like "racism" to understand racial disparities.[3]

One component distinguishes episode 3,697 from many other *Freedomain Radio* broadcasts: Molyneux's guest. Often, his interviewees fall well outside of the academic and political mainstream, with many episodes featuring White nationalists or conspiracy theorists.[4] In this particular episode, though, Molyneux is joined by someone inside the scientific establishment—Professor Stephen Hsu. Hsu is a theoretical physicist turned aspiring geneticist, who at the time was serving as senior vice president for Research and Innovation at Michigan State University (MSU). Raised in Iowa by Chinese immigrant parents, Hsu graduated from Cal Tech with his Bachelor of Science while still a teenager. YouTube brims with countless video interviews of Hsu, who enjoys nostalgically recounting anecdotes of his undergraduate interactions with the acclaimed physicist Richard Feynman (as he does in his conversation with Molyneux).

The Hsu episode of *Freedomain Radio* begins like any other academic interview. Molyneux introduces Hsu, describes his resume and expertise, and even praises him for founding two technology start-ups. Then, he asks Hsu to describe the current state of research into cognitive ability and discuss common misconceptions about IQ tests. Hsu explains that the effects of DNA play an important role in shaping an individual's cognitive test performance, using an anecdote about his older brother Michael. The MSU professor considers himself to be the smarter sibling, saying: "I think there's somewhat of a gap between [myself and my brother] in cognitive ability—no offense, Mike." Because the two brothers experienced a very similar upbringing (with the same family, gender identity, neighborhood exposures, etc.), Hsu attributes their differences to the different DNA each brother inherited from their parents. However, many factors other than DNA could explain different outcomes between two brothers, including different expectations from teachers, as the Rosenthal and Jacobson study from the 1960s demonstrated. The most rigorous genomic research on intelligence suggests that DNA explains some, but certainly not all, of the difference in intelligence between two brothers.

As the interview continues, Hsu and Molyneux seem to become more comfortable with each other, smiling and even laughing together. Around the halfway mark, their precarious conversation takes a sudden turn into more ethically perilous terrain. Molyneux mentions that they "must follow the facts, wherever they lie" and begins asking about genetic differences between racial groups. Hsu initially responds cautiously, saying that he has always been "agnostic on observed test score differences between [racial] groups . . . whether that's partially due to genetics." He concedes to Molyneux that he has "been agnostic, not because it's impossible, but because it's such a charged [topic]." Just minutes later, after some enticing by the *Freedomain Radio* host, Hsu obliquely states his belief on the issue, expressing his distaste for structural explanations of social inequality (a topic discussed further in the next chapter). Hsu says he doesn't believe that "there's some invisible miasma which is pushing, you know, Asian Americans up [and] African Americans down"; for him, this idea is "very ascientific."

Throughout the interview, Hsu nods along, even as Molyneux espouses dubious and ill-conceived views. Despite the two cities' similar sizes, why does Seoul have so much less crime than Mexico City, Molyneux asks? To Molyneux, the answer is obvious: the average IQ in Mexico hovers around "the sweet spot for criminality," whereas Seoul has a higher average IQ and therefore "criminality is just enormously lower."[5] In comparison to Molyneux, Hsu treads more carefully regarding the claims he makes about the world. As one would expect of a scientist, Hsu signals various issues where he thinks the evidence is stronger or weaker and, at times, even tries to articulate potential counterarguments to his views. Even so, throughout the entire hour-and-a-half-long interview, Hsu never meaningfully disagrees with or contradicts Molyneux. Hsu ends his interview by thanking Molyneux, saying: "It's been a pleasure, and I'd love to come back on the show."

While Hsu's responses may resemble a tacit endorsement, it's impossible to know whether and to what extent he shares Molyneux's beliefs; at a minimum, though, Hsu failed to adequately contradict the podcaster's racist, pseudoscientific ideas. The intersection of a mainstream academic scientist and a fringe White supremacist podcaster highlights how, while genetic myths are in part inherited from society's ugly eugenic past, they also have links to and can be influenced by present-day scientists, their research, and academic institutions. Scientific research can be communicated—and miscommunicated—in ways that buoy or undermine various genetic myths, like the Race Myth. The entanglement of genetic research and genetic myths, in turn, may have consequences for those outside of academia's ivory tower. Hsu's case illustrates the potential consequences even for those within the university ecosystem.

In 2020, a campaign led by MSU's graduate student union argued that, among other issues, Hsu was flirting with scientific racism, citing his online writings and interviews (including the one with Molyneux). A smattering of professors wrote letters in Hsu's defense, but the university's president soon asked Hsu to resign from his post as a senior vice president (and return to his non-administrative post as a tenured member of MSU's physics faculty). Since his demotion, Hsu has remained unapologetic, believing that those with "political and ideological motivations" unfairly attacked him for simply seeking to "preserve meritocracy and truth" within academia.[6]

Even when academic researchers do not directly converse with White supremacists, genomic research still gets picked up by those with ill intentions and used in contexts the researchers likely never imagined. Consider, for instance, genomic studies related to lactase persistence. Lactase is an enzyme crucial for the healthy digestion of milk. Usually, it is present in the body early in life and then disappears after weaning. However, roughly a third of humans maintain higher levels of lactase into adulthood—a trait known as lactase persistence. The majority of adult humans are *not* lactase persistent and will have difficulty digesting dairy products—the condition commonly known as lactose intolerance. Although many medical institutions conceptualize lactose intolerance as an illness, it is actually the norm for humanity.

Lactase persistence and lactose intolerance are influenced by DNA. The past decade of genomic research has found that rates of lactase persistence vary across different branches of the Family Tree, perhaps a result of divergent historical patterns related to the development of agriculture. Today, certain populations from East Africa, West Africa, Europe, and the Middle East tend to have the highest rates of lactase persistence. In a 2018 long-form piece, Amy Harmon, the Pulitzer Prize–winning journalist for the *New York Times*, described how White supremacists began appropriating recent genomic research on lactase persistence by selectively in-terpreting scientific studies and using them as additional proof of their racial superiority.[7] In effect, the White supremacists adopted lactase persistence as a stand-in for Whiteness itself, with genomic research serving as the connection between the two.

Snippets from academic papers on the genetics of lactase per-sistence began appearing on online message boards next to hateful arguments telling communities of color to leave America: "If you can't drink milk, you have to go back." The hashtag #MilkTwitter even went viral in 2017 when a group of White supremacists turned up at an anti-Trump art installation holding cartons of milk and began spewing racial, sexist, anti-Semitic, and homophobic slurs.[8] This "milk party" kicked off a slew of others like it: groups of White supremacists attempted to chug gallons of milk to prove their superior digestive capabilities. In fact, months before the

2022 massacre in Buffalo, New York, the shooter posted a selfie with a glass of milk on the instant messaging platform Discord. In the caption, he bragged that he could drink all the dairy he wanted, ending with "you're [*sic*] move brown boy." (Unfortunately for White supremacists, few people can chug an entire gallon of milk without vomiting or, at the very least, feeling deeply unwell.)

The Hsu-Molyneux and lactase persistence examples illustrate how genomic research can reinforce and promote genetic myths, regardless of the researchers' original intent or a study's actual conclusions.[9] This process, in turn, raises polarizing debate about the risks and benefits of genomics research, and especially genomics research on social and behavioral traits. History provides numerous examples of how claims regarding genetic differences in human behavior were used for social harm, but the unwieldy and expansive question of what the world today would look like without social genomic research does not have a clear-cut answer; there is no lab test or clever natural experiment to run. For this reason, disagreement about whether and how to conduct social genomics often stems from divergent fundamental beliefs about how the world and scientific institutions function.

———

Those in favor of social genomics research believe it will reap scientific benefits and tend to generally oppose limits placed on academic exploration. These enthusiasts are more likely to deemphasize the eugenic history associated with this type of science and instead focus their attention on the expansive set of possible futures. Could genomic data improve collective understandings of how well an intervention works—and if so, for whom? Even without a particular real-world application in mind, some motivate social genomics research using the key rationale underlying most basic science research: an increased understanding of the world is inherently valuable, even when the precise benefits are difficult to predict or quantify at present times. For example, the biochemical discovery that would eventually become a key piece of the revolutionary COVID-19 vaccines was largely ignored until the pandemic hit in 2020—even though it was developed years earlier.[10] Indeed, a lead author of the study recalled feeling disillusioned and downtrodden after it received five consecutive rejections from scientific journals, a signal that other academics did not see much value in his team's contribution.[11] Some scientific studies and findings do not appear to have a practical application until, suddenly, they do.

Over the past decade, useful scientific developments *have* come out of the field of social genomics—some of which could have been difficult to anticipate. Perhaps most importantly, such research has helped bring to the fore just how difficult it can be to distinguish causation from correlation when it comes to DNA. The genomic analysis of social traits has raised practical and conceptual challenges that also apply—but were nonetheless often overlooked—during the genomic analysis of medical traits.[12] Indeed, the genetic architecture of a range of medical traits, from depression to asthma to diabetes, overlaps with the genetic architecture of various social traits,[13] perhaps due to the fact that social factors often stratify an individual's exposure to health risks and their access to medical treatment.

A key insight from the last decade of social genomics research has been a greater appreciation of the intricate ways that people with different DNA are sorted—both by themselves and the institutions around them—into different social groups, schools, and geographic regions.[14] This complexity means that a person's genome contains granular information about their place within a broader population and that many large-scale genomic studies intending to identify the effects of DNA may simply be identifying the (nongenetic) effects of correlated environments. In addition, if a person has a given DNA variant, then so does at least one of that person's biological parents. Another important social genomic development shows that each DNA variant can affect a person twice: both indirectly, by first altering the phenotype of their parents (through a process sometimes referred to as "genetic nurture"),[15] and directly, when the person inherits the DNA variant. The first generation of genomic studies had no way of disentangling the effects of DNA from correlated environments or distinguishing genetic nurture from the direct effects of DNA; however, researchers are now beginning to shift toward using more flexible and robust within-family techniques.[16]

Moreover, proponents of social genomics tend to view scientific inquiry as a key tool for limiting the spread of ideology. If the Destiny Myth truly is a myth (as chapter 2 explains), then changes to the relationship between DNA and life outcomes across place and time should be observable. The field of social genomics has been at the forefront of the exploration of so-called gene-environment interactions:[17] how the social and physical environment modifies the effects of a person's DNA. Though there exist numerous recent studies that explore gene-environment interactions for social traits,[18] perhaps the most poignant example is one led by sociologist Pamela Herd, a professor of public policy at the University of Michigan.[19] Herd and her coauthors followed a cohort of Americans who were born around 1940 in Wisconsin—a social period when women in the United States were far more constrained in their ability to pursue higher education than they are today. When these Wisconsinites, who were largely of European ancestries, graduated from high school in the late 1950s, the educational attainment polygenic score was more highly associated with men's likelihood of attending college than women's. There's a kind of commonsense logic to this result: in the 1950s, most women couldn't go to college, regardless of how academically successful they were in high school. However, Herd and her colleagues found that as gender norms liberalized in the 1970s and 1980s, many of the women with high polygenic scores attended college later in life. The way each person's DNA ultimately manifested was dynamic and depended on key aspects of the social world. The findings from this gene-environment interaction literature emphasize that, for a wide range of traits, it is impossible to understand the effects of DNA without first examining the social conditions in which the DNA operates.

Other researchers feel that the juice simply isn't worth the squeeze—that the field of social genomics is, on balance, harmful. In previous eras, scientific research into the genetic underpinnings of social and behavioral outcomes helped to spawn genetic myths that, in turn, increased social inequality. Critics of social genomics tend to see history repeating itself—a world in which genetic myths continue to circulate, social inequality grows or persists, and science cannot cure ideology because science is an ideology itself, embedded in a broader unequal social and political system. The tendency to never stop asking questions drives science forward, but it can also distract from more pressing societal needs and give the impression that further study is required before taking effective action.

For some critics, the harms of genetic research may not be abstract possibilities because they may be members of communities that have been harmed by scientific research and genetic myths. To them, the risks may feel much closer to home, in their lives and the lives of their loved ones, than any scientific benefit seems to be. The excessive pursuit of research and data can delay action such that it never comes or comes too late. While figures such as Molyneux may seem like fringe cases operating on the margins of the media landscape, their words have power, and the scientific sheen of genomics research only enhances the sense that their beliefs are valid and worth acting upon. In the very worst cases, genomics research has played a role in precipitating truly shocking acts of violence.

———

The Kensington Expressway, a six-lane highway in Buffalo, New York, was constructed during the 1960s amid bitter controversy. Its development required the demolition of countless homes and businesses, as well as the razing of a tree-lined parkway designed by famed landscape architect Frederick Law Olmsted. The new expressway cut through an emerging middle-class Black neighborhood.[20] Perhaps unsurprisingly, property values in the area plummeted. To some, the ultimate consequence of the Kensington Expressway project—just like the many other US highway development projects in the works during the twentieth century—seemed to be limiting Black property ownership by cutting through predominantly Black neighborhoods and maintaining racial segregation.[21]

Katherine Massey—or Kat, as her family and friends called her—was among the Black residents who keenly felt the expressway's impact. She grew up on Cherry Street, which today runs parallel to the Kensington Expressway on Buffalo's East Side; Kat would spend the better part of her life living there. Unlike the expressway's likely impact on property values and neighborhood wealth, which is diffuse and difficult to quantify, the negative consequence that most concerned Kat was plainly visible to the naked eye: the Kensington Expressway had abruptly transformed Cherry Street from a quaint neighborhood avenue to what was essentially a highway frontage road. As a seemingly never-ending string of cars and trucks zipped down the expressway, Kat sat at home stewing in frustration. Much to her chagrin, the expressway was here to stay. Nevertheless, Kat, always strong in her convictions, decided she could still do something about it.

In the late 1980s, Kat took aim at the unsightly fence erected by the state Department of Transportation, which separated Cherry Street from the highway. Its railing had grown rusty, and the area behind it had become abandoned and overgrown. Kat drafted a letter to then-New-York-governor Mario Cuomo, explaining the situation and asking for his help. The state tried to give Kat the runaround, claiming that they only responded to requests from neighborhood associations or block clubs, not individual residents. Even in the face of this challenge, Kat was not one to give up easily, so she founded the Cherry Street Block Club.

Kat served as the block club's inaugural president, treasurer, secretary, and sole member. She sent another letter to the governor, now using official Cherry Street Block Club letterhead that she designed herself. This time, it worked. The Department of Transportation agreed to sand the rust off the railings and provide the materials needed to re-paint them. Armed with plenty of paint but no work crew, Kat enthusiastically recruited friends, family, and neighbors to get the job done.

Decades later, Kat would again work to give Cherry Street the facelift it deserved. The Department of Transportation had unveiled new plans to landscape the area between Cherry Street and the Kensington Expressway and replace the fence, which was in perpetual disrepair, with a more tasteful row of concrete dividers. Members of her community nominated Kat to provide input. At the first meeting she attended, Kat pulled out a piece of Kente cloth and requested that the designers integrate elements related to the neighborhood's African heritage into the new dividers. Kente cloth—brightly colored, woven fabric originally from Ghana—is popular among many members of the African diaspora who view it as a symbol of their shared ancestry.

The project moved forward at a languishingly slow pace, but, partly due to Kat's unrelenting calls to the Department of Transportation, it was finally completed in 2011. Kat first started her campaign to address the rusty railings on Cherry Street in the 1980s under Governor Mario Cuomo, and more than 20 years later, she finally got to relish the culmination of her hard work (during the governorship of Mario's son, Andrew Cuomo).

If you visit Cherry Street today, you will find concrete railings adorned with West African symbols and brief explanations of their meaning. On one such concrete slab, a symbol of a bird looking backwards appears next to the phrase Sankofa, which literally means to "go back and retrieve"; Sankofa signifies the eternal quest for wisdom and the need to reflect on the past in order to build a better future. Kat, with some help from her trusty Kente cloth, had once again improved her community. The very same year that the

Kat Massey

DOT completed the project, Kat retired from BlueCross BlueShield, where she had worked for 40 years.

In 2022, Kat was settling into her second decade of retirement and adjusting to life during the global COVID-19 pandemic. Every other week, she made a trip to the grocery store for meat, fruit, paper towels, and other essentials. Kat no longer drove, so she enlisted help from Warren, her younger brother, or Barbara, her younger sister. On Saturday, May 14, it was Warren's turn to serve as chauffeur. Usually, he would wait in the car while Kat did her shopping. That morning, something felt different—Kat asked her brother to head home and come back in 45 minutes. Kat climbed out of the car, and Warren pulled out of the Tops Supermarket parking lot. Tragically, it was the last time he would see his sister alive.

Kat Massey and nine others—Celestine Chaney, Roberta Drury, Andre Mackniel, Margus Morrison, Heyward Patterson, Aaron Salter, Geraldine Talley, Ruth Whitfield, and Pearl Young—were murdered by a White teenager who used the Race Myth as justification for committing a mass shooting.[22] A shocked scientific community would soon learn that the shooter's screed included screenshots, citations, and commentary on a number of peer-reviewed genomic studies, which he claimed as evidence for his White supremacist ideology.

———

The Social Science Genetic Association Consortium, or SSGAC, was founded in Boston in 2011. Its leaders, a group of economists with PhDs from top universities across the United States and Europe (including Harvard and MIT), were vexed by a scientific puzzle. Decades of research comparing identical and fraternal twins showed that a person's DNA sequence influenced a wide range of their life outcomes. While insights from twin studies indicated that DNA *in general* seemed to play a role, such approaches had no way of isolating *which* of the millions of DNA variants corresponded with any given trait. It was a frustrating time—imagine if a nutritionist told you that the type of food you ate played a role in shaping your cardiovascular health, but they couldn't identify which specific foods increased or decreased your likelihood of having a heart attack.

At the turn of the twenty-first century—just as the field of genetics was beginning to transform into the field of genomics—an innovative new set of technologies arrived on the scene. These technologies enabled the measurement of DNA variants across the entire genome which, in turn, allowed scientists to connect specific genetic regions with a person's life outcomes. The newly available genetic measures showed great promise; however, the field quickly ran into significant obstacles. A team of scientists would publish one result, and soon after, another group of researchers would come to an entirely different conclusion. Findings just could not be replicated. The SSGAC was founded on the idea that, if social genomics researchers wanted to make scientific progress, they had to change their approach. Each research study had too few people to rigorously run all the necessary scientific tests; the SSGAC argued that instead of competing with one another, scientists should pool their resources. Through mass collaboration, they could make steady, robust, and replicable scientific progress. The first figure in this book (a scatterplot in chapter 1 featuring polygenic scores generated by the SSGAC) illustrates the resulting advances in polygenic prediction.

Like most academic researchers, the members of the SSGAC juggle a host of responsibilities on any given day. They might draft responses about their in-progress article to peer reviewers; coordinate statistical analyses using more than one hundred genetic datasets (containing individuals spanning from America to Estonia to Australia); organize a small army of research assistants, many of whom hope to obtain a coveted letter of reference for graduate school; write FAQs to help journalists communicate the results of social genomics findings to broader audiences;[23] plan workshops to train the next generation of social genomics researchers;[24] or, finally, obtain public and private grant funding to keep the titanic research machine afloat.

In 2018, the SSGAC published the third in a series of scientific studies on DNA and educational attainment. The study, informally referred to as "EA3," utilized the largest sample ever analyzed in social genomics thus far—over *a million* individuals. Interestingly, a few years later, James Lee, one of the study's lead authors, was one of the academics who sent a letter to MSU's president urging against Stephen Hsu's removal. Lee wrote that "a malicious and ill-informed Twitter mob has congealed over the last few days, calling for Hsu's removal . . . however, you must not surrender a single inch to it."[25] (Lee is no longer a member of the SSGAC.)

Those outside of the ivory tower hardly notice the release of most academic papers, but EA3 made quite a splash. The *New York Times* covered the study's findings in an article titled "Years of Education Influenced by Genetic Makeup, Enormous Study Finds."[26] A piece entitled "How Scientists Are Learning to Predict

Your Future with Your Genes"[27] appeared in *Vox*. Pulitzer Prize–winning journalist Ed Yong brought the study's findings to the public in his *Atlantic* article: "An Enormous Study of the Genes Related to Staying in School."[28]

EA3 was not only picked up by mainstream media outlets. White supremacist groups on online forums and messaging platforms like 4chan and Discord also took interest. Just weeks after the study was published, a pseudo-scientific table appeared online that cross-referenced the EA3 results with data from a different genetic study, known as the 1000 Genomes Project,[29] that explored geographic variation in DNA across the globe. The table deceptively used statistical artifacts to "prove" that White people have a genetic intellectual advantage over Black people. The 4chan user who produced the table wrote that "the data is pretty damning." In reality, EA3 had nothing to do with differences in intelligence across races. Still, the table went viral; it has since appeared thousands of times on 4chan.[30]

Soon enough, in May 2022, EA3 appeared in mainstream media once again. The headlines read differently this time: "Buffalo shooting ignites a debate over the role of genomic researchers in white supremacist ideology" ran in *STAT News*.[31] "Science Must Not Be Used to Foster White Supremacy" appeared in *Scientific American*.[32] *Undark* magazine published a multipage spread titled, "A Field at a Crossroads: Genetics and Racial Mythmaking."[33] EA3, along with a number of other genomic studies, had appeared online in the Buffalo shooter's screed as supposed proof of genetic differences in intelligence between White people and Black people.

Many of the specific claims made in the Buffalo shooter's screed were unoriginal; he largely borrowed arguments, propped up by distorted science, that had been making the rounds online among White supremacists for years. Two-thirds of the images included in his screed were taken from 4chan, and more than three-fourths of the rationale section came from online hate websites.[34] The link between these arguments and the shooter's violent acts nonetheless raised the alarm for many scientists and funders, who wondered how mainstream science had become a citation in the rantings of a scientific racist. Recently, population geneticist Je-didiah Carlson and his team showed that the online dissemination of pseudo-scientific tables and images, like those in the Buffalo shooter's manifesto, has been steadily increasing since 2016.[35] In the wake of the Buffalo shooting, many in the scientific community became concerned about how social genomics research could be responsibly conducted—with some even wondering if it should be conducted at all.

Debates about the risks and benefits of social genomics rage on, but few stop to ask who the debaters are. Who is setting the research agenda, and who gets to decide how to best conduct this research? Whose values and experiences are being factored in when determining potential harms and benefits? The stakes of social genomic research are high. They are higher now than ever before, and they are high *for everyone*. Still, the scientific machine marches on—with very few given a say.

Academic researchers—people like Daphne and Sam—make up a tiny fraction of the US population. Importantly, the data are very clear: academics are not representative of the general public. Of course, academics complete more years of schooling than the average American; they are also more likely to have had a privileged socioeconomic upbringing. Tenure-track faculty tend to grow up in wealthier neighborhoods, and they are nearly twenty times more likely than the average American to have a parent with a PhD.[36] Academia is something of a family business; Sam's father, for instance, is a professor, and now Sam is, too. Significant racial disparities exist in academia, as well. Even though 32% of Americans identify as either Black, Hispanic, or Indigenous, just 13% of all professors are members of one of these three groups. These disparities become more pronounced toward the top of the academic ladder and are typically even more extreme in STEM-related fields, including human genomics.[37]

Even academic researchers with the very best of intentions tend to have limited sets of knowledge and lived experiences. The day-to-day life of a member of the SSGAC is very different from the day-to-day life of someone like Kat Massey, who exists outside of the field of academic study. Even if the profession became more racially and socioeconomically diverse, academics would still be academics—focused on publishing scientific research, teaching, and securing funding. Yet, one way or another, academic or not, research into the human genome affects everyone.

The development of new technology often requires striking a delicate balance between encouraging scientific advancement and controlling the spread of potential harms. In rare cases, the harms and benefits of a technology fall at the two extremes. These dual-use technologies, as they are called, present both immensely generative and extraordinarily destructive outcomes, and special restrictions must be put in place to try to minimize harm. Nuclear technologies serve as a classic example: they can provide clean energy to power entire cities, but they can also produce weapons that could reduce those same cities to ash. Similarly, research on viruses that may help prevent the next pandemic also poses the threat of biological terrorism. Today, some argue that advances in artificial intelligence, which have already produced elegant solutions to stubborn scientific problems (like mapping the intricate structure of proteins),[38] threaten to upend civilization as we know it.

Social genomics research does not meet the traditional definition of a dual-use technology. Although it cannot build bombs, such research can still cause significant harm—and this "weaponization" features prominently in debates about whether and how the results of genomic studies should be regulated. (Part 3 discusses regulation and public policy in more detail.) The difficult, controversial questions about whether and how to conduct social genomics research will only grow. As polygenic scores are developed for a wider range of traits and as existing polygenic scores become more accurate, it is important to consider who should have access to which genomic information. Since the launch of the Human Genome Project, the idea that insights derived from the human genome should benefit all of humanity has fostered a relatively open and democratic scientific environment, where genomic discoveries are shared freely and widely. Regulating access to, for example, the DNA information used to construct polygenic scores—what researchers call GWAS summary statistics—may stand in opposition to open science; however, it may at some point play a necessary role in combatting irresponsible, inappropriate, or socially harmful applications.

In 2019, for example, a startup company used the summary statistics from a recently published genomic study on same-sex sexual behavior to release an app titled "How Gay Are You?" For around $5, consumers could upload their DNA data and received their results for a same-sex sexual behavior polygenic score.[39] "How Gay Are You?" triggered alarms within the scientific community, including among the authors of the original genomic study, who stressed that their results could *not* be used to accurately predict a person's sexuality. Today, in a range of countries around the world, members of the LGBTQ+ community can face the death penalty. Could results from the "How Gay Are You?" app be used as evidence for such a sentence? The company eventually pulled the app from the market, but as this book covers in later chapters, there exist plenty of other questionable direct-to-consumer genetic tests based on GWAS summary statistics.

There is an urgent need for a society-wide conversation about what is next for genomics research and the resulting genomic technologies. Navigating the tough questions at hand will be possible only by considering and including the Kat Masseys of the world—those who stand to experience the most harm. Everyday people are notably absent from the scientific research establishment's business as usual, even though US taxpayers fund a substantial proportion of research in the country.[40] However, a growing number of efforts are underway that aim to give voices to groups historically excluded from the scientific research process.

———

In late 2019, bioethicists Erik Parens and Michelle Meyer teamed up to facilitate an academic discussion on the risks and benefits of social genomic research. They brought together academic researchers from many different disciplines (including Daphne!) who spent three years reflecting, discussing, and often outright arguing about the issue. Despite their fair share of disagreements, the group agreed that genomic research on traits and outcomes related to social status (e.g., educational attainment), tied to marginalized identities (e.g., sexual orientation), or historically used to perpetuate stereotypes (e.g., intelligence) requires a heightened commitment to responsible research conduct and communication. The group also agreed that genomic research involving these sensitive traits and outcomes that compares racial, ethnic, or ancestral groups is, at present, scientifically invalid. In a world where such research might become feasible, they determined, it should not be conducted, funded, or published without a compelling scientific and/ or ethical justification.[41]

The Parens-and-Meyer-led initiative began as an academic exercise in considering the risks and potential benefits of social genomics. After recognizing the need to include the perspectives of members of the public, they decided to recruit a community sounding board partway through; Daphne spearheaded this effort, bringing together a small group of individuals from across the United States and intentionally recruiting people from communities historically harmed by scientific research (e.g., those with lower household incomes, lower levels of education, and who self-identify as racial or ethnic minorities).

Over the course of 18 months, the sounding board met online via Zoom to learn about and discuss social genomics. Community members identified many of the same risks and potential benefits that researchers had, but they may have been weighing those factors differently from researchers. For instance, members of the sounding board seemed especially concerned about how data generated by social genomics research might be used by law enforcement in ways that could directly harm them.[42] One member shared during a meeting: "I and many people from the African immigrant community—we, just like other people of color in this country—we sometimes get worried about how research data is used by law enforcement to stereotype the whole population." Another member voiced a related concern, asking: "If you find that [a] certain type of people have the warrior gene . . . [that] they are more prone to aggression or they're more prone to steal or whatever—do people use that information, then, to restrict them?"

Sounding board members also shared their concerns about how private industry might utilize genomic data. Several mentioned that they thought industry is driven solely by profit rather than by a desire to help people. One member talked about Henrietta Lacks, a Black woman from Baltimore whose cervical cancer cells were taken without her consent in the 1950s and became widely used in biomedical research. Remarking on Henrietta's story, she said: "Lots of companies benefitted and made lots of money from herself, and still do today." Another sounding board member shared—in reference to polygenic scores—that he believed "We live in a very capitalistic society—someone is going to profit off of this."

Community sounding boards like the one Daphne facilitated illustrate just one approach for creating a more inclusive scientific research process. Members of the public also volunteer their time to review NASA data in the search for planets in other solar systems—an example of citizen science.[43] Indigenous scientists and tribal members are partnering to train the next generation of scientists and ethicists; to help ensure that Indigenous peoples shape their own research agendas, they are also creating their own facilities to collect biological samples.[44] There are increasing calls for scientific frameworks that include patients and community members on research teams rather than merely as objects of study.[45] For instance, in 2014, after the water crisis broke out in Flint, Michigan, the city's residents partnered with Michigan universities to found two centers: the Healthy Flint Research Coordinating Center and the Flint Center for Health Equity Solutions.[46] (The Flint water crisis is an issue close to Sam's heart: his PhD dissertation focused on understanding the educational impacts of the crisis on the affected children.) These centers were established to ensure that the needs of Flint's residents were prioritized over scientists' research agendas.[47]

Unfortunately, these examples largely remain the exceptions that prove the rule. Many people, both within and outside of the scientific community, believe that this trend should change.[48] Genetic epidemiologist Genevieve Wojcik, for instance, has argued that the genetics community can take a stronger stand against harmful genetic myths by, among other things, increasing engagement with the public.[49] When asked to reflect on their participation, one sounding board member shared: "When your research has broad effects that anyone in the global population can be harmed or helped by . . . taking into account the opinions of nonscientists is essential." The future of genomics research depends on identifying the best approaches for broadening who in society has a voice in conversations about how it designed, conducted, and communicated.

———

SAM: *Ultimately, while research on social genomics can be misused, I think that—on balance—the benefits outweigh the risks.*

DAPHNE: *I can appreciate the potential benefits of research in social genomics, but all told, I fall on the other side of the spectrum—from my perspective, the risks are likely to outweigh benefits.* The benefits are, at least at this point in time, more diffuse, conceptual, and upstream in the scientific process, whereas the risks are more immediate and high-stakes.

SAM: I truly don't know how this current round of research on DNA and human behavior will go. I can absolutely see the potential for misuse or misunderstanding of studies in social genomics. Still, I don't think we're doomed to reinforce myths about genes—the past is not destined to become the future. After all, is this not the same kind of determinism that is so problematic about the Destiny Myth?

DAPHNE: My so-called determinism comes from the fact that very little meaningful or positive social change is happening *in the world right now—change that I think is necessary for creating a world where social genomics is not implicated in serious harms, like a racially motivated mass shooting. It's not impossible; I just don't see signs that it's happening. Instead, I see how social genomics research is breathing new life into genetic myths. Plus, plenty of people still seem to completely ignore the fact that science doesn't operate in a vacuum and can create social harm just as it can social benefit!*

SAM: *Even if you don't think studying how and why DNA variants are related to various social and behavioral traits is worth the risk, it's not clear to me how exactly we would feasibly study some traits but not others. The first hard question is where and how to draw the line between what is social and what is medical. Even once that's done, DNA affects us through the physical and social world that we have collectively built. So, because a person's social characteristics— including their education—stratify their access to healthcare in America today, when we study the genomics of various health outcomes, we will accidentally pick up the effects of DNA on education. In a sense, it's all social genomics research—and studying the genetics of various diseases without any knowledge about the genetics of education is kind of like juggling with one hand tied behind your back.*

DAPHNE: *To be honest, it still makes me uncomfortable to hear you say "the genetics of education"—and we've been talking about social genomics for over five years at this point! People often ask me why I study the ethical and social implications of social genomics and spend so much time talking to social genomics researchers. I think that some of them worry that, in doing so, I may help legitimize research that they view as fundamentally flawed and harmful. Still, I believe there's more harm in pretending that this research does not exist or writing it off. This research isn't likely to stop, and overstated claims about genes that have created real harms for real people aren't simply going to disappear. Also, very few policies and incentives exist to consider the downstream effects of not only social genomics but scientific research more broadly. I want to live in a world where we seriously consider and work together to prevent these downstream harms. So, I cannot imagine how we could accomplish that without paying attention to this research, trying to mitigate against its risks, and better understanding the motivations of those who conduct it— people like you, Sam!*

SAM: *Well, we agree that genes can play an outsized role in our social imagination—that merely acknowledging that DNA influences behavior can reawaken long-standing myths about genes. Inside the research establishment and out, we all have responsibilities to safeguard against the risks social genomics research may introduce.*

DAPHNE: *True, and including the voices of people like Kat Massey in decision-making processes about social genomics is an important start. I'd also like to see additional guidelines from funders and academic journal editors on how to conduct social genomics research and report its findings. We also need policymakers to start thinking about how these research products get marketed and sold to consumers—from social traits to medical traits to everything in between. At a basic level, I think more restrictions are needed to make it harder for these data to find their way into the hands of those with ill intent. It's possible to allow good science to proceed while also mitigating risks.*

SAM: *Hold on, we're getting ahead of ourselves! Before talking about regulation and policy, we need to talk a bit more about DNA and social inequality.*

6

DNA and Social Inequality

DAPHNE: *Not long after we first started collaborating, I learned that you had donated one of your kidneys to someone you had never met. We'd been working on an academic paper together, but I had no idea you were going through this process until your PhD advisor mentioned it to me. It was wild. I remember thinking: "Who is this guy? What motivates someone to donate an organ to a total stranger?" It struck me as both selfless and, to be honest, hard to comprehend—and it was through talking about your kidney donation that I got my first insight into how you think about social inequality.*

SAM: *You aren't alone—my parents also found it a bit hard to understand, especially at first! I had to explain to them that, for a young donor like me, the risks are minimal, and the benefits to the recipient are enormous, freeing them from endless hours of dialysis and adding about ten years to their life.[1] Many folks say they believe in helping those in need, but I often feel frustrated at the general lack of urgency—how little people actually put their money where their mouth is. Charity isn't really charity unless it costs you something. From our early conversations, I thought you and I were on the same page in terms of how we thought about social inequality—but then Kathryn Harden's book* The Genetic Lottery *came out, where she argues that DNA matters for understanding and addressing social inequality.[2]*

DAPHNE: *Wow, we had such different responses to reading that book! At first, I didn't get why we reacted so differently. We both want to build a world where the opportunities afforded to the most fortunate members of society are similar to the opportunities afforded to the least fortunate. I had a tough time figuring out why you see genomic research as relevant to conversations on social inequality when I do not.*

SAM: *Our distinct understandings of social inequality led to some of our most difficult conversations. Without realizing it, we had very different notions of what social inequality is at its core. We were using the same term to refer to two very different ideas.*

DAPHNE: *Yeah, I have a distinct memory of us working in my kitchen in San Francisco. We had pulled out a whiteboard and were trying to chart the different ways people might come to define social inequality and how these distinct definitions either make a person's DNA relevant or not. The whiteboard kept getting messier and messier, and we just could not understand each other. We ended up having to take a break and come back to it the next day.*

SAM: *I'm glad we stuck with it, though. As we found out, reasonable people can agree that social inequality exists but disagree about whether DNA can help us understand and, perhaps, address it—it all depends on what a person means by "social inequality."*

———

Cornell Charles—or "Dickey," as he was affectionately known—was born and raised in New Orleans, Louisiana. At the age of fifty-one, he still lived in the house he grew up in. Dickey and his wife, Nicole, had an open-door policy. On Sundays after church, their house would fill up with friends and family who arrived—both invited and uninvited—in search of a home-cooked meal. Dickey always delivered: heaping bowls of gumbo, plates of spaghetti salad, miniature meatloaves. He was also a proud member of the Zulu Club, a Black community organization known for its lively Mardi Gras parade, scholarships, and food collection programs. One evening, not too long after Mardi Gras, Dickey returned home from work and told his wife he didn't feel well. The next day, he spiked a fever. About a week later, on March 24, 2020, Dickey drew his last breath—becoming one of over a million Americans to die from the COVID-19 virus.[3]

The unprecedented conditions brought on by the introduction and spread of COVID-19 took a toll on all Americans, but the early days of the pandemic most acutely affected Black and other racial minority communities.[4] In Philadelphia, where Sam lives, nearly 50% of deaths from COVID-19 occurred within the Black community, which makes up less than 40% of the city's population.[5] Across the country in San Francisco, which Daphne calls home, the Hispanic population comprised more than 25% of COVID-19 deaths despite representing only 15% of the city's residents.[6] Black, Hispanic, and Indigenous Americans were roughly twice as likely to be hospitalized and die from COVID-19 in 2020 and 2021 than White Americans,[7] in part because they are more likely to be uninsured,[8] suffer from health conditions like diabetes or hypertension that amplify complications from COVID-19,[9] and live in multigenerational households (increasing the risk of transmission).

COVID-19 laid bare the many processes that come together to produce social inequalities across the globe. Differences in quality of life between individuals—from access to food, water, and shelter, to the threat of violence, to educational and economic opportunity—are a pervasive feature of the world today. While many in the wealthiest parts of the world enjoy extravagant amenities, over 700 million people (more than one-tenth of the world's population) live in extreme poverty, subsisting on less than $2 per day.[10] Social inequality not only exists among people who live in different regions and countries but also among neighbors dwelling in the very same community. Whether in a country, city, or neighborhood, countless factors—including race and class—stratify people, shaping the life opportunities afforded some over others. In the United States, those at the top of the income distribution can expect to live more than a decade longer than those at the bottom.[11] Taking just a cursory look, it is impossible to ignore that many folks are constrained in their ability to live life to the fullest and, sometimes, to even live a good life at all.

The distribution of adversity and opportunity in society does not result from mere chance. Instead, an intricate web of social, economic, and political processes produces and recreates inequality across place and time. Trade deals benefit some and harm others; a domestic factory relocates overseas, propelling one region into industrial decline and prompting economic growth in another. Families transmit advantage and disadvantage across generations; educated, wealthier parents move to certain neighborhoods to ensure their children go to the very best schools, while less educated, poorer parents struggle to put food on the table. Patterns of racial segregation stratify access to healthcare; Black Americans in large urban areas are more likely to live farther away from life-saving medical care.[12] Laws allocate who has rights and, perhaps more importantly, who does not; a woman who immigrated to the United States from one country may be granted asylum, whereas a woman who immigrated from a different country is deported.

Many people hope to help build a more socially equal world. Researchers study social inequality with the belief that a thorough understanding of the production of inequality will strengthen efforts to address it.[13] Some of them argue in favor of focusing on understanding and addressing structural inequality, centering disparities such as racial gaps in college completion or socioeconomic gaps in access to healthcare. Now that scientists are starting to measure and observe genetic risk, others argue for a more expansive view of social inequality—one that also includes the adversity or advantage that DNA differences between individuals may produce. Perhaps unsurprisingly, the idea that DNA differences between individuals are a source of inequality has proven to be highly contentious.

As it turns out, disagreements about DNA and social inequality often boil down to far-reaching disagreements about what exactly social inequality is and, consequently, to what extent addressing structural inequalities between groups should take priority over other sources of difference. Some processes, like redlining, produce differences in well-being across key facets of society: Black people and White people, rich Americans and poor Americans. Other processes, like the age at which a person enters school and, perhaps, DNA, shape a person's life in ways that do not cut cleanly across groups. Are these also sources of inequality? (Note that reducing inequality is not society's *only* priority but one important goal to be balanced among others—like increasing average levels of human flourishing.)

How a person defines social inequality is central to whether they think DNA is relevant to efforts to reduce it. As chapters 7 and 8 cover, different conceptualizations of social inequality are also crucial for considering the risks and benefits of polygenic scores (and, by extension, their regulation). Before starting to think about the incoming tide of DNA-based technologies, it's necessary to first dive into the ongoing debate about DNA and social inequality.

———

Structural inequality, as used in this book, is an umbrella term. It may include, for example, structural racism (sometimes also referred to as institutional racism) or structural ableism. It can be applied to specific contexts, such as education or the economy, or more broadly to society. The specific structural categories that are advantaged or disadvantaged by a given social structure may vary across place and time.

Structural inequality exists as the result of unjust human actions—public policies and social institutions that systematically afford some groups of people privileges, resources, and opportunities that others are denied.[14] For example, involuntary sterilization laws disproportionally targeted impoverished women (like Carrie Buck) and women of color, thereby diminishing their power. A structural approach emphasizes that social inequality and injustice do not exist merely because of the behavior of a few specific "bad actors"; rather, systems like racism and classism are embedded into the very fabric of a society, from laws to institutions to culture. Those who take a structural view therefore tend to believe that building a more socially equal world will require a comprehensive overhaul of the current organization of society.[15]

Demonstrating structural inequality doesn't require a deep dive; it appears all around us, if you know where to look. In 1984, James Fletcher, a fifty-six-year-old Black man who laid railroad tracks across the American South, ran into a problem. The Atlanta house he owned and lived in needed a new roof, but he was having trouble receiving a construction loan. James and his wife, Lizzie Mae, went to Citizens & Southern, their bank of ten years. However, the bank promptly turned them down: "They said they didn't make no house loans," James recounted in an interview. "They didn't let us fill out the papers."[16]

Eventually, the couple found a different lender, Atlantic Mortgage, willing to give them a loan. However, because their loan was deemed risky, James and Lizzie Mae had to pay an effective interest rate of a whopping 27%. This meant that the total amount they would end up paying for their $5,773 loan was over $30,000. If Citizens & Southern had offered them the loan at their usual terms, James and Lizzie Mae would have instead owed less than $12,000. Referring to Atlantic Mortgage's exorbitant interest rate, James said: "I don't know exactly how the thing went, but . . . I guess they [the bank] can get by with it." James and Lizzie Mae were powerless; with so few options to choose from, they could either accept the unreasonable offer from Atlantic Mortgage or return home to a roof growing leakier by the day.[17]

Investigative reporter Bill Dedman told James and Lizzie Mae's story in the *Color of Money*—a 1989 Pulitzer Prize–winning series

James Fletcher

for *The Atlanta Journal*.[18] The set of articles explores redlining, a discriminatory practice in which financial services, like a home loan, are withheld from potential customers who live in neighborhoods classified as "hazardous" to investment. Dedman's reporting showed that Atlanta banks would often lend in lower-income White neighborhoods, but not in middle- or even upper-income Black neighborhoods.

Redlining is an example of a process that produces structural inequality. It came to exist not by accident but instead by careful design. In the aftermath of the Great Depression, the federal government undertook dramatic reforms to stabilize the housing market. One such reform created the Home Owners Loan Corporation (HOLC), an agency that drew colorful maps of more than two hundred American cities to document the "riskiness" of lending in various neighborhoods. Neighborhoods were classified based on housing age, quality, and price, as well as characteristics such as the race, ethnicity, and immigration status of the neighborhood's residents. Since the lowest rated neighborhoods—which typically housed Black residents—were drawn using the color red, these maps gave rise to the term "redlining."[19]

One HOLC technical document describes how a neighborhood in Tacoma, Washington, might have been classified as yellow instead of red "if not for the presence of the number of Negroes and low-class Foreign families." Similarly, in Youngstown, Ohio, a red area is attributed to the "ever growing influx of Negroes and low-class Jewish."[20] While redlining became illegal in 1968 with the passage of the Fair Housing Act, its impact on American cities endures; to this day, areas that had more redlining in 1930 have elevated rates of racial segregation. In addition, persistent racial gaps remain in rates of mortgage approval.[21]

In being denied a construction loan, James and Lizzie Mae were victims of structural inequality. Redlining played a key role in producing and reproducing disparities between Black and White Americans in a variety of valued social and economic life outcomes.[22] James and Lizzie Mae had to pay a very high interest rate on the loan they eventually received because the loan, or perhaps Black people and majority-Black neighborhoods, were deemed "risky." Unable to access credit at a fair price, the couple accrued less wealth to pass onto their children, just like countless others in similar situations.

A common strategy for quantifying structural inequality is to measure systematic disparities across certain social and demographic characteristics (like race). The enduring legacy of redlining is illustrated in the fact that, today, wealth disparities between White and both Black and Hispanic families continue to be substantial. White families have more than five times the average wealth of Black and Hispanic families. In 2019, though the median White family had wealth valuing nearly $200,000, the median Black family and Hispanic family had wealth valuing just $24,000 and $36,000, respectively.[23]

This book largely focuses on racial and economic disparities (because, past and present, these forms of structural inequality are often invoked in genetic myths). However, many of those who take a structural view of social inequality also include other "structural categories"—such as gender identity, sexual orientation, language, religion, citizenship, and disability. Structural categories are social and demographic characteristics that shape how an individual is treated by laws, culture, and institutions. Such categories are also often attached to ideologies of inferiority and superiority. Sometimes, people disagree about which categories should be viewed as "structural" in a given place and time, with debates often pitting individuals experiencing different forms of marginalization against each other and ultimately serving to maintain the status quo. If you ask some people, like Daphne, what social inequality is, they describe structural inequality; for them, the two words are synonymous. However, for other people, like Sam, there are additional sources of inequality to address—sources that disparities across common structural categories do not capture. Note, while some thinkers distinguish between undeserved and deserved inequality, this chapter is largely agnostic to such a distinction and applies to any formulation of deservingness.[24]

Some events cause people's life experiences and outcomes to diverge but lack a clear structural explanation: why does one passenger suffer a permanent neurological injury after a car accident, while their friend in the next seat is entirely unscathed? Why is one town pummeled by a tornado while a neighboring town is miraculously left untouched? Such events do not cleanly correspond to unjust institutions or individual actions, and they do not cut cleanly across social lines. Still, these events, as well as the unique string of As, Cs, Ts, and Gs (DNA nucleobases, discussed in chapter 1) that each person inherits from their parents, cannot be anticipated or controlled. At the end of the day, people differ in their

What *is*
social inequality?

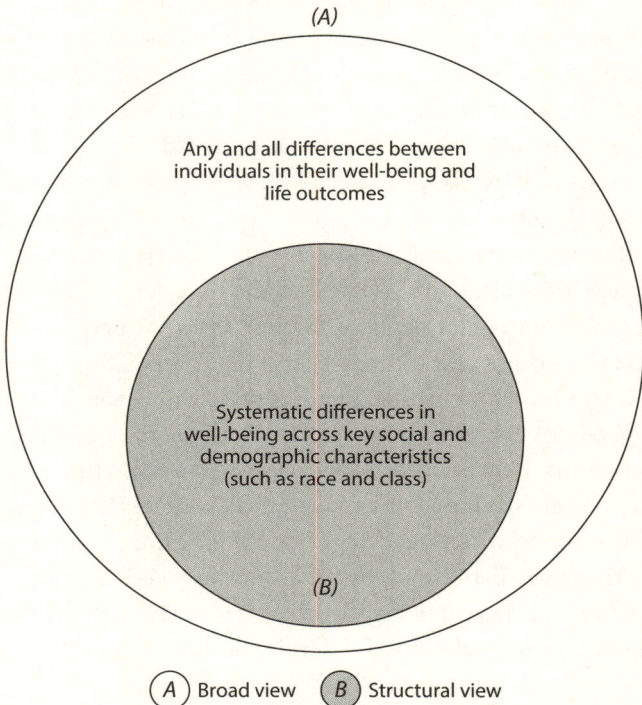

(A)

Any and all differences between
individuals in their well-being and
life outcomes

Systematic differences in
well-being across key social and
demographic characteristics
(such as race and class)

(B)

(A) Broad view   (B) Structural view

FIGURE 5. Differing conceptualizations of social inequality

amounts of well-being. Those who take a broader view of social
inequality often believe that—regardless of whether those differences result from an oppressive system, discriminatory actions, or
a random occurrence (like the DNA they inherit)—everyone deserves the chance to have an authentic, meaningful, and happy life.
Building a more socially equal world therefore requires seeking to
reduce *any and all* differences in well-being between individuals.

The two conceptualizations of social inequality presented
in figure 5 differ in terms of which sources of difference each
focus on. A structural view of inequality, represented in the

shaded portion of the figure, centers differences between specific groups. A broader view of inequality, represented in the non-shaded portion, steps back to look at *any* differences between any individuals.

———

Siblings often share similarities; Sam and his older brother, Max, for instance, have a lot in common. They share the surname Trejo, the same two parents, and mixed Mexican and Ashkenazi Jewish heritage. They also share memories of singing along to the Beatles with their dad and the unique experience of having first names that seem like diminutives but that, in fact, are not (their birth certificates say Max and Sam, not Maxwell or Samuel). The brothers both grew up in Austin, Texas—back when the city was still actually a little weird—and attended the same magnet school.

So far, Max's and Sam's life outcomes are pretty similar. They are both generally healthy college graduates with white-collar jobs that put them toward the top of the US income distribution. Neither of them has been arrested, and both tend to vote for similar candidates in elections, despite their shared mild disillusionment with the political system. Max and Sam love craft beer, and they both sometimes feel anxious in social situations.

The fact that siblings frequently share early life experiences and adult outcomes is not surprising. Siblings raised in the same family tend to share many structural categories, such as race, class, culture, religion, and citizenship. An abundance of social science research highlights how families transmit cultural and social capital to their children, which, in turn, shapes their interests, aspirations, and opportunities.

Max's and Sam's similarities reflect a broader trend. Researchers have used nationally representative samples of Americans to explore the extent to which siblings turn out similarly or differently. Their results show that a little over half of the variation between adults in important life outcomes, such as years of schooling, income, and health, is explained by factors that siblings share.[25] These shared factors include family income, race, neighborhood, and local school quality, among others. In short, the family into which a child is born plays a key role in shaping their educational, economic, and health outcomes, thereby impacting their overall well-being. On average, adversity in childhood begets adversity in adulthood; similarly, privilege in childhood begets privilege in adulthood. Intergenerational mobility—the extent to which individuals are, in practice, able to escape the circumstances of their birth—is relatively low in the United States, so there is still plenty of work to do to reduce this inequity.

But while the family and community characteristics that are shared by siblings shape a person's well-being, a sizable portion of the variation still remains unexplained. Even siblings can end up living markedly different lives. In the case of Sam and his brother, some of these differences are relatively trivial in the grand scheme of things: Sam earned a doctorate, whereas Max didn't pursue schooling beyond receiving his bachelor's degree. Max is a vegetarian, while Sam is a meat-eater. Sam lives on the East Coast, and Max lives on the West Coast. Sometimes, though, sibling differences are anything but trivial.

In his first year of graduate school, Sam sprained his ankle while playing a pickup game of ultimate frisbee. Max, a lifelong soccer player, had experienced and successfully recovered from similar sprains countless times—but for some reason, Sam's injury affected him permanently. Doctors would discover that Sam had damaged his tibial nerve, which would constantly send faulty pain signals to his brain, leaving him unable to comfortably stand or walk. The resulting chronic pain would consume his life for years—nothing, even multiple invasive surgeries, seemed to help. Eventually, Sam discovered a way to manage his pain using a new medical device—one that may as well be straight out of a sci-fi novel. He has a surgically implanted wire in his calf and wears a battery that sends electrical pulses to it, scattering the pain signals before they reach his brain. Sam's chronic pain will likely always be with him, but this nerve stimulator helps to quiet it.

Sometimes, Sam cannot help but wonder: why did such a minor injury produce such a catastrophic result? Why didn't any of Max's many ankle sprains leave lasting damage? Perhaps the different strands of DNA each Trejo boy inherited from their parents played a role. It's possible that Sam's unique DNA sequence put him at higher risk for neurological damage. The existence of meaningful differences in well-being between siblings raises a key question: when the life outcomes of people born into very similar social and economic circumstances diverge, do these differences contribute to social inequality?

What if a person believed that *any and all differences* between individuals in their well-being and life outcomes constitute social inequality? This view on social inequality includes structural inequalities like the residential segregation and unfair lending practices of redlining that James and Lizzie Mae experienced. It also includes, for example, differences in the lifespan between siblings that result from their DNA, as well as differences in peoples' well-being that result from factors as arbitrary as the cutoff date for kindergarten enrollment.

Across the world, birthday thresholds govern when kids start school. Children are born at all times of year, but society often lumps them into distinct cohorts. For example, New York City has an enrollment cutoff date of December 31, meaning that any child who will turn five years old by the end of the calendar year must be enrolled into kindergarten. Thus, the kindergarten class for 2026–2027 in the nation's largest school district will be made up of children born January through December of 2021. Texas, where Max and Sam grew up, has a slightly different cutoff date: September 1. Max, who was born in July, entered kindergarten at age five and tended to be among the youngest in his class. Sam, on the other hand, was born in February. He entered kindergarten at age five and a half, placing him right in the center of his grade's age distribution. The cutoff date that a school district sets for enrollment may seem harmless, but many studies show that the age at which a child enters kindergarten actually matters quite a bit.[26]

Many parents believe that having a child older than the rest of the class is beneficial. This belief is reflected in the relatively common practice of "red-shirting," in which a parent delays a child's entry to school.[27] The idea behind red-shirting is that children who have been allowed to mature for another year will benefit more from their schooling. Unfortunately, little rigorous evidence supports this perception.[28] It is true that, in any grade, older children tend to perform better academically than younger children (that is, six-and-a-half-year-old children perform better in school than six-year-old children on average). However, all else equal, starting school earlier is actually beneficial.

Imagine two very similar children: one is born the day before their state's age cutoff, and the other is born the day after. Using this natural experiment, researchers have found that school enrollment cutoffs, while seemingly insignificant, have a meaningful impact on life outcomes. On average, entering school later decreases a child's tests scores and their likelihood of graduating from high school, depresses their adult earnings, and increases their chances of being incarcerated.[29] Kids who start school at younger ages benefit from having older peers (for example, think about a freshman in high school developing soccer skills more quickly by practicing with the varsity team). These children have additional time in school as a child and in the labor force as an adult. The main downside to entering kindergarten earlier is an increased rate of ADHD diagnosis, thought to be driven by the fact that younger children are developmentally more immature compared to their older classmates.[30] Over a person's lifetime, these small advantages and disadvantages compound on themselves, producing meaningful differences in adulthood that are a result of school-starting age.

So, while structural categories like race and class shape opportunity and affect individual differences in well-being, other processes—even ones as trivial as the timing of one's birth relative to an arbitrary cutoff date—also end up mattering for a person's life outcomes.[31] Importantly, there are key differences between processes, like a child's school-starting age and the processes that produce structural inequalities. For one, no set of ideologies in the United States suggests a social hierarchy of birthdays. People born at one time of the year are not viewed as superior to those born at different times; indeed, the fact that school-starting age cutoffs vary across place and time suggests that there is no sense of universally "better" or "worse" birthdays. (On the other hand, there are ideologies used to suggest and justify class- and race-based hierarchies.)

Additionally, the differences in well-being among individuals that result from school-starting age are not accumulated and reproduced in the same way that structural inequalities often are—parents do not typically pass on their birth date to their children, and children born into rich versus poor families, or Hispanic versus Asian families, are not systematically born at different times of the year.[32] In short, school-starting age produces real differences in well-being between individuals—differences that some folks argue are worth noting. However, the differences in life outcomes produced by school-starting age do not fit easily into a structural understanding of social inequality.[33] While structural inequality focuses on differences that result from unjust institutions, systematic oppression, and discrimination, a broader view of social inequality will also pay attention to differences in individuals' well-being that stem from less systematic occurrences.

———

These different conceptions of social inequality lie at the heart of the debate about DNA and social inequality. Early genetic thinkers used their discoveries about the inheritance of DNA to promote eugenic policies and justify social inequality. Some worry about a similar story playing out with modern genomic research, especially for studies that focus on social and behavioral outcomes.

In *The Bell Curve*, published prior to the completion of the Human Genome Project, Richard Herrnstein and Charles Murray mounted a dubious argument that social inequality in the United States was genetic in nature. Using no DNA data whatsoever, they purported American society had become genetically stratified: that a wealthy, so-called cognitive elite had grown genetically distinct from the rest of society. The pair also controversially argued that racial disparities in a range of life outcomes stemmed from differences in DNA. Following the DNA revolution, as genetic data has become increasingly available to researchers, many of the empirical claims made by Herrnstein and Murray have been proven false.[34] Still, decades later in 2020, Murray wrote an op-ed asserting that the DNA revolution offers the potential to validate his and Herrnstein's initially unfounded claim, arguing that polygenic scores are "impervious to racism and other forms of prejudice."[35]

Murray's attempt to divorce the innate influence of DNA from environmental processes is inherently fraught; it stinks of the Destiny Myth. In a response to Murray, three members of the SSGAC pointed out that polygenic scores, as well as the effects of DNA more broadly, "can and do reflect racism, sexism, or other prejudices, as well as more benign environmental factors."[36] If children lack access to healthy meals or educational enrichment because of their skin color, a polygenic score for educational attainment (and most other traits for that matter) would reflect this inequality. Similarly, if a child's dyslexia goes undiagnosed and causes them to repeat a grade because the school they attend lacks adequate funding for special education, a polygenic score for educational attainment will capture this tangled result. Our profoundly unequal world distorts and refracts the effects of inheriting various strands of DNA. In other words, the effects of DNA are not innate and cannot be understood in absence of the accompanying environmental factors. Inequalities between racial and class groups exist *because* of discrimination and unequal opportunities. Those who use DNA to explain such differences are merely putting up a corrosive smokescreen.

Many researchers work hard to avoid Murray's pitfalls, and some even hope to shape a new narrative about DNA and social inequality—one that recognizes the falsity of the Destiny Myth and the fact that polygenic scores capture long and complex chains of cause and effect (that scholars, at present, hardly understand themselves). These researchers argue that precisely *because* the effects of DNA are not immutable, there exists a moral imperative to reduce the extent to which DNA shapes well-being. In this way, polygenic scores should not be used as tools to validate or explain social inequality but rather as tools to bolster efforts to enhance social equality.[37] Still, as sharp critiques of Paige Harden's book *The Genetic Lottery* demonstrate (including one written by Daphne),[38] such views have drawn the ire of many progressive scholars and policymakers.[39] How is it that different progressive thinkers, who agree about the existence of social inequality and a collective duty to reduce it, disagree about the role of DNA?

Crucially, different answers to this question depend on what people mean by social inequality. Do they prioritize the view that social inequality is the result of differences in well-being between key social and demographic groups—that is, structural inequality? Or do they tend to believe that social inequality is what happens when there are any differences between individuals (irrespective of the source of the differences)?

As genomic data and research continue to proliferate, researchers continue learning more about the ways in which an individual's DNA can affect their well-being. The DNA that a person inherits can help explain why biological siblings end up with different educational attainments, disease diagnoses, or even lifespans.[40] Some researchers contend that without DNA data, some of these sibling differences may go unnoticed (or noticed too late for optimal intervention). Over the past two decades, though, these previously invisible genomic processes have started to reveal themselves. For this reason, some consider DNA and genomics research relevant to conversations about social inequality; they argue that in order to identify and monitor inequality that stems from DNA differences, tools like polygenic scores are a necessity (similar to how racial demographic information is required to monitor racial inequality).[41] However, for those who prioritize structural inequality (rather than all forms of difference), polygenic scores have little to offer.

Structural categories came to exist because of specific human-driven processes (e.g., colonialism and slavery). Sure, DNA variants may produce differences in skin tone upon which racial discrimination is targeted, like in the case of Millie and Marcia (discussed in chapter 2). Still, polygenic scores are not necessary for witnessing or identifying such racism—studies of racial discrimination have long preceded the availability of genomic data.[42] Similarly, understanding why American children from upper-class families test roughly three grades ahead of children from working-class families does not require analyzing DNA[43]—simply studying the different experiences of such children at home and at school may provide those answers.[44] There is no scientific evidence to support the idea that race and class disparities, the structural inequalities that we focus on in this book, are the inevitable result of DNA differences across groups. In short, researchers and policymakers interested in structural inequality do not need DNA to achieve their goals.

While the DNA revolution is rapidly giving researchers the capacity to measure a wider and wider range of individual differences, polygenic scores offer little upside and a number of potential downsides for those concerned with structural inequality. Integrating DNA into the broader project of measuring social inequality may not only breathe new life into genetic myths but also distract from how the current social arrangement produces and/or maintains inequality. As the sociologist Ruha Benjamin points out, tractable social reforms already exist; an expansive literature documents how poverty negatively impacts a person's educational prospects. Claims about the need for more data and research on threats to health and well-being (and on which people these threats affect most) risk justifying a failure to collectively act. This "datafication of injustice" lets people off the hook for their lack of action.[45] To reference the sociologist Matt Desmond, "We don't need to outsmart this problem. We need to outhate it."[46]

Even those who take a broader view of social inequality and are optimistic about the use of polygenic scores to build a more equal world must take this concern seriously. Although polygenic scores may allow us to begin measuring and documenting the previously hidden genetic characteristics that shape our lives, these new tools force us to face difficult questions. How many societal undertakings can reformers simultaneously make progress on? Does having the goal of building a society where every person has the ability to flourish—regardless of the DNA that they inherit—exist in alignment, or in tension, with efforts to address long-visible racial and socioeconomic inequalities? Integrating new characteristics into the broader project of measuring social stratification could split people's attention, ultimately drawing away from existing efforts to reduce the most urgent and entrenched inequalities.

One fact, however, is clear: when it comes to conversations about DNA and social inequality, people often ask and answer very different questions. Is social inequality synonymous with structural inequality? Is social inequality something more expansive? Definitions aside, those who *can* work to reduce social inequality *should*. Building a society that is more compassionate, just, and fair is neither simple nor straightforward, and rhetoric must be backed up with concrete action. People can vehemently disagree about whether DNA helps or hinders the ability to rise to this challenge. Regardless, those who want to create a more equal future should agree on the need for greater regulation of DNA-based tools like polygenic scores—because without them, structural inequalities are likely to grow.

———

DAPHNE: *It wasn't until we started talking that I consciously realized I hold a structural view of social inequality. I mean, I knew I did, but I suppose I didn't appreciate the fact that other people, especially others concerned about social inequality, might think about it differently. I'm drawn to the structural approach because it brings attention to the issues that I think really matter, like housing discrimination, educational inequality, and how institutions are designed to conserve the power of some while simultaneously oppress others.* **I think society should prioritize the reduction of structural** *inequality over other kinds of inequality. For me, social genomics research is, at best, agnostic towards efforts to reduce structural inequality and, at worst, antagonistic to these efforts.*

SAM: *I've realized that I hold a broader view of social inequality. I do agree, though, that a structural lens for thinking about inequality is useful and important; understanding the institutions and social hierarchies that undergird inequality must inform potential solutions. Just because rebuilding the very structure of our society is difficult does not excuse us from trying. But at the end of the day, what matters most to me are people's lived experiences—the collection of moments that make life feel rich and moving.* **I believe that reducing differences between individuals in terms of their well-being creates a more equal world—regardless of which differences are reduced.** *If one day the test score gap between rich and poor students is narrower than it is today, we've reduced social inequality. Similarly, if one day disparities between those with high and low genetic risk for depression are lessened, we've decreased social inequality.* **So, I can see how social genomics could be used toward efforts to reduce (broad) inequality and have a positive impact on people's lives.**

DAPHNE: *Still, I worry that focusing on DNA could excuse us from trying to address (structural) inequalities—or at least distract us from them! I can't wrap my head around the argument that DNA is relevant to social inequality because, at least to me, social inequality is not some abstract idea to debate and theorize about. It is a concept defined by and tied to the urgent realities of our current world—and the specific things we can do right now to help those who need it. In my work as a bioethicist, I see every day how the amount of care and attention a person receives from our healthcare system depends on their social class and their race. Polygenic scores can't help us reduce those inequalities.*

SAM: *I mean, I think about that all the time. I know there are countless other people, in the United States and around the world, who are in pain and could benefit from the same medical device I have implanted in my leg—they just don't have the access. I still find it scary to think of how my life would have turned out without the top-notch medical care that my attachment to elite universities affords me.*

DAPHNE: *Then, what draws you to a broader conceptualization of social inequality?*

SAM: *Well, I think it's not always easy to distinguish which inequalities should count as structural or not. Events often result from many intertwined factors, ranging from unjust actions to random luck and everything in between! I believe we can reduce social inequality without agreeing on the "right" structural categories. If historians in a few hundred years look back on life today, which structural categories will we have missed? Recent social science studies have highlighted that a person's life outcomes remain largely unpredictable, which encourages humility regarding our current understanding of the factors that matter most.*[47]

DAPHNE: *In a different world—one with far less structural inequality and without harmful genetic myths—I can buy the argument that we should explore how DNA differences between individuals affect life outcomes. If we eliminated structural inequalities, then sure, we should be working hard to understand whether, how, and to what extent DNA differences explain why one woman lives longer than her sister. The truth is, though, that we don't live in a world without structural inequality! That was my issue with the* Genetic Lottery: *I felt like it was depicting an impossible world where we can effectively address racial and class inequalities while at the same time reducing the impact of the DNA one inherits. Our resources are finite. It is a no-brainer to me that we should be devoting those resources to reducing entrenched structural inequalities like poverty. Zooming in on DNA will not help* us get there. *The world we're living in right now needs our attention.*

SAM: *We may disagree about the extent to which DNA is a part of the production of social inequality, but I think there's something related that we agree on. Research, including studying the effects of DNA, is only useful to the extent that we do not understand inequality well enough—but at times this doesn't seem to be the key problem. In the United States today, we have policies that could cut child poverty in half or provide life-saving medicine to more Americans, but somehow the political will is just not there. While genomics studies may improve our conceptual understanding of the various processes that produce social inequality, I am somewhat skeptical that knowledge will, in turn, improve the lives of the less fortunate. Maybe an egalitarian country like Norway could consider using DNA to measure and reduce previously invisible forms of inequality, but the United States might just not be ready.*

DAPHNE: *Totally—I think the real problem is that we aren't serious about reducing inequality in the United States, not that we don't know how. That's why it's so important to take the risks of genetic myths seriously. It feels downright ahistorical to me when people argue that today's genomics research is going to suddenly dispel genetic myths—especially social genomics research. Many people still believe that social inequalities are the result of innate differences instead of the decisions we as a society make over and over again about how to treat one another.*

SAM: *We spent the better part of a year writing just this one chapter. Even after countless hours arguing, I don't think our respective beliefs about social inequality have fundamentally changed. Do you think that's okay? Has anything good come of all of our difficult conversations?*

DAPHNE: *I think so! We have come out of those discussions with better understandings of each other and of this issue more broadly. Someone can define social inequality differently than*

I do but still be motivated to address and prevent it—and I think recognizing that shared motivation can take us a long way.

SAM: *The absolute best course I took in college was by a philosopher named Paul Woodruff. During the first lecture, he told all the students that—with any luck—we'd all know a lot less by the end of the semester. Of course, Professor Woodruff didn't want us to literally lose knowledge. Instead, he wanted us to better recognize the limits of our understanding, to become intellectually humble and realize that many of the concepts that we thought were neat and tidy are actually messy and uncertain. After working with you, especially on this chapter, I know a lot less than I did before. And that's a good thing.*

DAPHNE: *I love that, and I agree with you. Throughout the process of writing this book, there have been so many moments where I felt lost and unsure . . . Unsure about what precisely my argument is, and then lost about how to effectively communicate it to you. We've talked a lot about the allure of simplification—and also about who benefits when complex ideas are made to seem straightforward. I think it's almost always the people who are doing the simplification, the ones shaping the narrative, who benefit. Ultimately, this chapter exists because our many arguments helped us to become suspicious of clean, simple answers when it comes to the questions about DNA and social inequality. Was leaning into the complexity uncomfortable? Yes. Was it ultimately worth it? For sure.*

PART III

Looking to the Future

7

Polygenic Prediction at the Fertility Clinic

The 1997 science fiction film *Gattaca* depicts a dystopic future in which children are conceived in the lab to ensure they possess the "best" genetic traits possible—for instance, decreased disease risk, enhanced athletic ability, and protection against addiction. Little is left to chance. Fertility specialists utilize a procedure known as *genetic embryo selection* to help ensure that offspring inherit the very best genetic traits of their parents. When one couple expresses reluctance to use the procedure for their future child, a clinician reassures them: "Keep in mind, this child is still you— simply the best of you. You could conceive naturally a thousand times and never get such a result."[1] He adds that, for an additional fee, a different procedure known as *gene editing* can allow the deletion of undesirable strands of DNA from an embryo's genome or the insertion of DNA not present in either parent.

In *Gattaca*'s geno-hierarchy, individuals conceived via genetic selection and/or gene editing are considered "valid." Anyone conceived naturally, the result of a so-called faith birth, is deemed "invalid." The leading roles in *Gattaca* were played by Ethan Hawke and Uma Thurman (who, as it happens, got together on set and had two children of their own). Hawke portrays Vincent Freeman, an invalid with poor eyesight and a high genetic risk for heart disease. Although the elder sibling, Vincent grows up living in the perpetual shadow of his valid younger brother, Anton. At every turn, Vincent is fed a dystopian version of the Destiny Myth, told that he will never achieve personal and professional success because of his inferior DNA: "I belonged to a new underclass, no longer determined by social status or the color of your skin," Vincent narrates. "No, we now have discrimination down to a science."[2]

To moviegoers in the late '90s, the dystopia portrayed in *Gattaca* may have seemed like mere science fiction (emphasis on the "fiction" aspect). However, the two technologies depicted in the film—genetic selection and gene editing—are scientific realities today. In November 2018, the scientific community was shocked by a thirty-six-year-old Chinese researcher's announcement on YouTube. Wearing a blue button-down shirt with a lab bench half-blurred in the background behind him, He Jiankui casually revealed that the world's first gene-edited babies had been born. The twin girls "came crying into the world as healthy as any other babies," the man who covertly engineered their conception declared proudly.[3] The sisters, Nana and Lulu, were now at home with their parents, Mark and Grace (He used pseudonyms for privacy). Gene editing, He explained, would help countless children avoid lifetimes of suffering.

He, who holds a PhD in physics from Rice University and completed a research fellowship in the Department of Bioengineering at Stanford University, had done what seemed to many ethically unthinkable. While a professor at China's Southern University of Science and Technology, He had used a recently developed gene editing technology called CRISPR to modify human embryos. In particular, he modified the DNA of embryos to resist HIV infection by disabling the CCR5 gene, which allows HIV to infect human cells. Notably, while He collected sperm cells from an HIV positive male, the disease cannot be transmitted from parent to embryo via sperm cells; the embryos that He had genetically modified were never actually at risk of contracting HIV. While He argued that the father's HIV status qualified as an unmet medical need, the scientific community strongly dissented. When two of those embryos became Nana and Lulu, the human germ line was irrevocably altered: the twins can pass the modification on to any of their own future offspring.[4]

By conducting his experiment, He had committed a number of research violations, including failing to obtain the proper ethical approvals. His use of CRISPR was also in clear defiance of a global consensus that more research was needed before gene editing human embryos.[5] When the news first broke, Chinese state-run media celebrated He's work, but global backlash against the scientist's unprecedented use of this new and uncertain technology swiftly followed. Facing mounting criticism, the Chinese government reversed course. He's research was suspended, and he was indicted alongside two collaborators. The once rising star within China's scientific research establishment was found guilty of research misconduct, sentenced to three years in prison, and forced to pay fines exceeding $400,000. The Chinese government amended its criminal code to ban the implantation of gene-edited human embryos, and an international group of researchers called for a global moratorium on all clinical uses of human gene editing.[6] The World Health Organization also stepped in, stating that any clinical applications of human germline editing would be, at this time, irresponsible.[7]

As news about He's action began to spread, researchers around the world voiced their concerns about the health and well-being not only of Nana and Lulu, but also of Amy—a third baby that He later admitted to having gene-edited. Very little is understood about the downstream effects of gene editing; there is a possibility that Nana, Lulu, and Amy (and any of their future offspring who inherit their edited genes) will experience unanticipated adverse effects later in life. Some research has found that disrupting CCR5—the gene He disabled to resist HIV infection—adversely affects bone growth.[8] Other evidence suggests that editing CCR5 could enhance memory and cognition.[9] In fact, at least one prominent stem-cell scientist suspects that He used CCR5's role in HIV to mask his true intention: cognitive enhancement.[10]

Many believe that the gene editing He so brazenly carried out has the potential to fundamentally reshape humanity—and not necessarily for the better. While scientists and policymakers are guarding the front gate against gene editing, genetic embryo selection (using polygenic scores) is slipping in through the back door. Even as the world came together to denounce He's use of gene editing, the very next year, an embryo was selected using polygenic scores. Hardly a squeak of dissent was heard. There is no global moratorium on embryo selection using polygenic scores. In the United States, there's virtually no regulation of the practice whatsoever. So, what is polygenic embryo selection, and how does it work? What are the social and ethical implications of this newly emerging reproductive technology? Moreover, what will it take to, at minimum, prevent it from exacerbating existing structural inequalities?

When the movie *Gattaca* was first released, in vitro fertilization (IVF) technology was still a relatively new process. Louise Brown, the first baby conceived using IVF, was born in northwest England in 1978, just two decades prior to the film's release. Her birth—like the birth of Nana and Lulu nearly half a century later—inspired immense controversy. Many considered lab-based conception perverse and immoral. As a form of hate mail, people sent the Brown family test tubes full of fake blood. IVF became a topic of national conversation, and a camera crew was stationed outside the hospital during Louise's delivery.[11]

A few years after Louise Brown made her entrance, the first American "test tube baby," Elizabeth Carr, was born. After what the Brown family had gone through, Elizabeth's parents feared harassment. They worked to elude the public eye, at times even assuming false names.[12] When Elizabeth was born healthy, the family's identity became public. The Carrs agreed to speak with the press, and a photo of an eight-month-old Elizabeth was emblazoned on the cover of *Life* magazine.[13] Years later, during an interview at age seventeen, Elizabeth described some of the stigmatization she had experienced from others: "They expect me to have thorns growing out of somewhere or that I'm psychic . . . but I always say, 'No, I don't think I'm any different.'"[14]

While many protested against IVF in its early days, the procedure has since become widely accepted. Today, IVF is increasingly common and has helped to conceive over eight million babies worldwide—including Millie and Marcia Biggs, the mixed-race twins introduced in chapter 2. Now, only one in eight Americans report having a moral issue with the procedure, and seven in ten support access to IVF (as well as a majority of Americans who are opposed to abortion).[15] Even as the political and legal climate surrounding reproductive rights and technologies evolves, the American public continues to support access to IVF. When the US Supreme Court overturned *Roe v. Wade* in June 2023, it opened the door to legal challenges against IVF. Subsequently, in 2024, Alabama's highest court ruled that frozen embryos were legally considered children under existing state law, forcing many of the state's IVF clinics to halt operations practically overnight. However, Alabama's legal limitations on IVF prompted widespread public outcry, leading state lawmakers to move quickly to pass legislation protecting IVF providers from legal liability.[16]

As the world moves deeper into the genomic era, the application of IVF technologies is only becoming more complicated. In fact, polygenic scores are starting to transform the reproductive landscape and, in turn, shape the next generation of human beings. The onset of this wave of new technologies comes with collective moral and policy questions. Consider, for example: should prospective parents be allowed to choose which embryos to implant using polygenic scores, thereby selecting the genetic characteristics of their future offspring? What could go wrong if only some people are given access to this technology?

While the impacts of polygenic score applications are uncertain, the potential stakes are high. The policy recommendations outlined in this chapter and the next balance innovation with caution; they are intended to prevent harm, especially with regard to the widening of structural inequalities.[17] It will take time to learn about the various ways that polygenic scores can be applied in society; the policy frameworks suggested in these chapters enable applications of polygenic scores to proceed slowly, introducing and adjusting regulatory policies incrementally as knowledge about this new technology—and how to use it—develops.

———

In 2019, Thuy Phan, an IT support worker, and Rafal Smigrodzki, a practicing neurologist, wanted to have a baby. Both Thuy and Rafal were in their fifties, so IVF using an egg donor and surrogate was their best option. Rafal, who has some training in genetics, had read a magazine article about a newly available option for those undergoing IVF: polygenic embryo selection.[18] The couple chose a recently established New Jersey–based biotech company called Genomic Prediction to conduct the necessary polygenic testing. Genomic Prediction was co-founded in 2017 by none other than Stephen Hsu—the ousted Michigan State administrator who was a convivial guest on a White supremacist's podcast (Hsu appears in chapter 5). Leading the company's commercial wing was none other than Elizabeth Carr—America's first IVF baby.[19]

Thuy and Rafal needed to find a fertility clinic willing to collaborate with Genomic Prediction, but many in the local medical community were uncomfortable with utilizing a technology that was still in its infancy. When the couple asked a physician in their hometown of Charlotte, North Carolina to help them use polygenic embryo selection, the clinician firmly refused, calling the practice unethical. Rafal was baffled; in his mind, it was a no-brainer to use what he considered to be a life-saving technology.

The two were not easily discouraged. Eventually, they found a different doctor—this time in Washington, DC—who agreed to work with them. The couple produced thirty-three embryos, but only four were deemed viable for implantation. The lab report that Thuy and Rafal received from Genomic Prediction for each of these four embryos included polygenic scores for conditions like diabetes, skin cancer, high blood pressure, and elevated cholesterol. The couple ultimately decided to implant the embryo with the lowest polygenic score for heart disease. In the summer of 2020, their baby girl was born. Elated, Thuy and Rafal named her Aurea—the Latin word for "golden." Their radiant little girl was the first child known to be born as a result of polygenic embryo selection.

The exact same technology that allows prospective parents to select embryos with low polygenic scores for heart disease can also be used to select embryos with high polygenic scores for more controversial traits, like intelligence, height, and physical attractiveness. At least as recently as 2020, Genomic Prediction advertised polygenic testing for intellectual disability and idiopathic short stature.[20] These terms—which simply refer to individuals whose IQ or height, respectively, falls below a certain threshold—allow companies to covertly market polygenic embryo selection for intelligence and height to prospective parents. The potential to select embryos using polygenic scores for traits like intelligence is, in part, what left Thuy and Rafal's first physician in North Carolina uneasy. Rafal, a self-described "techno-optimist" who hopes to have his brain cryogenically preserved after his death, believes that polygenic embryo selection will eventually follow the path of IVF and become less controversial.[21]

In a 2021 marketing video released by Genomic Prediction, Rafal—dressed in a white lab coat—enthusiastically describes polygenic embryo selection as "giving [a child] the best genes you can."[22] He characterizes it as "a truly revolutionary technology," with beneficial impacts that will likely surpass vaccines and sanitation. He confidently concludes that polygenic embryo selection "most likely will make a big difference in [Aurea's] life."

For now, Genomic Prediction has stopped advertising polygenic testing for intelligence and height. Hsu explained that although the technology is available, the company no longer offers this information "just because it's too controversial."[23] Even so, he has publicly speculated about one day offering polygenic embryo selection for skin color.[24] While nearly 60% of US adults disapprove of using polygenic embryo selection to select skin color, around 20% said that they approve.[25]

Since its founding, Genomic Prediction has been joined by a host of other companies. Some of them, like Orchid Health,[26] are openly selling polygenic testing for intellectual disability. In 2024, investigative reporting found that another company called Heliospect Genomics was charging couples up to $50,000 to test embryos for intelligence, claiming that their methods could result in a gain of more than six IQ points.[27]

While currently just one in sixty, or a little under 2%, of American babies are conceived using IVF, that figure may rise as infertility rates continue to climb.[28] In addition, if polygenic embryo selection grows in popularity, it is possible that some prospective parents may begin using IVF solely for the opportunity to influence the genetic characteristics of their future child. Rafal certainly hopes polygenic embryo selection will become "the default method of making babies."[29]

Even if many clinicians feel uneasy about polygenic embryo selection,[30] roughly 70% of Americans approve of it.[31] If provided for free, nearly 40% of Americans say they would utilize polygenic embryo selection to increase their child's likelihood of getting into a selective college.[32] At the same time, however, around 30% of Americans think that the use of polygenic embryo selection should not be permitted at all.[33]

Currently, there are no legal restrictions in the United States on which traits prospective parents can target using polygenic embryo selection. An urgent society-wide conversation is needed to develop regulations that get ahead of this rapidly evolving technology, including ethical considerations that guide which traits—if any—are acceptable to select for. Before wading into the waters of morality, though, the state of the science needs evaluating; notably, the scientific limitations of current polygenic scores shape what is even, at present, feasible. How accurate are existing polygenic scores (and for which traits)? How well can scientists guess a person's eventual traits using only embryonic DNA?

———

Frank Gilbreth Jr. and his sister Ernestine Gilbreth had an unconventional childhood. They grew up in a family of an astounding *twelve* biological siblings—six boys and six girls. In a semi-autobiographical novel, Frank Jr. and Ernestine offer a comical account of their unusual upbringing during the early twentieth century in Montclair, New Jersey. They describe how their father, Frank Sr., was often asked about his and his wife Lilian's decision to have so many children. In response to the question, Frank Sr. would cheekily reply: "Well, they come cheaper by the dozen!"[34] The Gilbreth family (and Frank Jr. and Ernestine's novel) served as inspiration for the numerous iterations of the comedy film *Cheaper by the Dozen*.

Though most couples end up having only one or two children, considering couples that have *many* children, like the Gilbreths, is useful for understanding the range of possible characteristics that can be passed on from parents to their children. Imagine that Frank Sr. and Lillian were of average height for men and women, respectively. Height is a polygenic trait that can be easily measured and observed and has been of keen interest in studies of genetic inheritance at least as far back as Francis Galton (introduced in chapter 2). If Frank Sr. was about 5′10″ and Lillian was around 5′4″ (the average height for each sex in the United States),[35] their sons would grow up to be, on average, about 5′10″, and their daughters would grow up to be, on average, around 5′4″. This is just *on average*—in actuality, pairs of brothers and pairs of sisters do not typically turn out to be exactly the same height as each other. (Daphne and her three sisters, for example, range from 5′6″ to 6′0″.) Sometimes, due to sheer chance, the genetic material that one son inherits from his parents has a greater number of height-increasing (versus -decreasing) DNA variants than another son receives. Alternatively, perhaps a daughter is born during a tough economic year for the family and does not receive quite enough nutrients while in the womb, making her shorter than her sister, who was in utero when times were better. For these reasons, some of the Gilbreth brothers may turn out to be 6'0", but others may only grow to be 5'8". Some Gilbreth sisters may reach 5'6", whereas others barely push 5'2".

The dozen Gilbreths each began as an embryo that was conceived naturally when the first of Frank Sr.'s sperm cells reached an awaiting egg cell in Lillian's womb. Except for the case of twins, natural conceptions only ever involve a single fertilized embryo. However, conceiving via IVF works differently. In IVF, a doctor collects sperm and egg cells from the parents-to-be. Then, in the lab, the doctor combines sperm and egg to create numerous fertilized embryos. How do people then choose which embryo to implant?[36] Among any set of so-called sibling embryos, there are differences in terms of which genes each inherits from their parents. Consider how the heights and personalities of siblings in any family typically vary; for example, while Ernestine Gilbreth was quite adventurous, often getting into mischief with her brothers, her younger sister Martha was far more studious, preferring to stay in and read a good book. Embryo selection is a procedure that leverages genetic differences between sibling embryos to choose and implant an embryo that is more likely to develop certain attributes, thereby influencing the characteristics of the eventual child.

When IVF was first introduced in the late 1970s, doctors selected an embryo to implant using far less information than they have available today. In effect, the selection process was quite similar to conceiving a child the traditional way. Starting in the 1990s, however, the ability arose to select an embryo based on sex and to determine whether it had specific monogenic diseases. American parents could now specify whether they preferred a male or female,[37] and special care was taken to make sure not to implant embryos that were expected to develop known single-gene disorders, like cystic fibrosis.[38] In 2019, with the implantation of the embryo that became Aurea, decisions regarding which viable embryo to implant started to be made using polygenic scores.

So, how exactly does polygenic embryo selection work? If Frank Sr. and Lillian were alive today and elected to use IVF instead of having a litter of children the old-fashioned way, their fertility doctor would first collect sperm cells from Frank Sr. and eggs cells from Lillian. Imagine that they develop around a dozen viable embryos.[39] Even if the couple decided that they wanted a male and screened for monogenic diseases, they would still be left with, say, six embryos to potentially implant. Then, Frank Sr. and Lillian would be offered the option to use polygenic embryo selection. If the couple agrees and decides to select on height (they want their son to be as tall as possible), how well would polygenic embryo selection for height work? Researchers like to think about the impact of the procedure in terms of the *expected change*—so, in the case of height, how much taller a child resulting from polygenic embryo selection is expected to be than a child conceived without it.

Imagine that the six Gilbreth brothers were magically cast back into the embryo stage of development. Each so-called sibling embryo is deposited in its own petri dish, and the six embryos are placed side-by-side in a fertility lab. Encapsulated within each one is a unique DNA sequence, as every sibling embryo inherited different strands of DNA from Frank Sr. and Lillian. Unlike gene editing, which would introduce novel DNA variants into the sibling embryos, polygenic selection leverages only the DNA variants that a given set of parents has. The goal of polygenic embryo selection would be to guess which embryo, if implanted, would grow up to be the tallest Gilbreth son.

Exactly how effective a particular polygenic score is at this task depends on a few factors. First, there is the extent to which DNA is implicated in the underlying causal processes of the trait being selected for (i.e., heritability). For instance, while a person's DNA strongly influences their height, it hardly affects the intensity of their religious beliefs—otherwise known as religiosity.[40] Therefore, it likely will never be possible to meaningfully increase a child's religiosity using polygenic embryo selection. However, in contexts where consistent and quality nutrition is widely available, height can be especially influenced by such a procedure. From a practical standpoint, polygenic embryo selection for height will be more effective than for religiosity.

The second factor that shapes how useful a particular polygenic score is for embryo selection pertains to the specific research study used to generate the score. The number of genotyped people included in a study is vital to the accuracy of a polygenic score. In addition, how well researchers can account for the effects of study participants' demographic characteristics (like where a person lives and how wealthy their family is) also matters. Due to the sheer size of the human genome, genomic studies need many, many participants to identify precisely which genetic variants correspond to slight increases or decreases in a given trait. If the sample size of a genomic study is too small, the polygenic score created from that study is highly imprecise; a brother with a high polygenic score and a brother with a low polygenic score will not actually have very different heights. When a polygenic score is imprecise, clinicians cannot reliably identify the embryo most likely to grow the tallest or the one with the lowest risk for a particular disease.

Geneticists can make an educated guess as to how accurate the polygenic score for height will be in the long run. The best theoretically possible polygenic score for height is expected to produce a change of 2½ inches when selecting from ten embryos. (These estimates depend, in part, on the number of viable embryos—the more embryos to choose from, the wider the range of genetic characteristics between them, and the greater the expected change.) In 2010, when researchers published one of the very first modern genomic studies of height—with a sample size of nearly 200,000 individuals—the resulting polygenic score produced an expected change of just over ½ an inch.

In 2022, a little more than a decade later, researchers amassed an astounding sample of 5 million individuals—the largest used in a genomic study yet—to search for DNA variants related to height. The resulting polygenic score produced an expected change of 2 inches—close to the theoretical maximum of 2 ½ inches. Importantly, however, this expected change is only currently possible for embryos whose location in the Family Tree is near the 5 million individuals studied, who largely hail from Northern Europe. (The later parts of this chapter discuss the inaccuracy of polygenic embryo selection for ancestries that are underrepresented in genomic studies in more detail.) An expected gain of 2 inches is not nothing, but it is also not huge.[41] Selecting for multiple different traits simultaneously (for instance, height *and* heart disease) will typically reduce the expected gain of each trait. In other words, researchers would currently only expect a 2-inch change when selecting for height *on its own*.

For most traits, the accuracy of polygenic scores lags well behind that of height. For example, Thuy and Rafal used the polygenic score for heart disease to select the embryo that became their daughter Aurea. Unlike height—where someone can be 4′10″, 6′7″, or anything in between and which forms a classic bell-shaped distribution—a clinical diagnosis of heart disease is a condition that someone either has or does not have. If a person has enough buildup of plaque in their arteries to meet the clinical definition, they are diagnosed with heart disease. Even though heart disease is a condition that you either do or do not have, scientists prefer to conceptualize *risk* for heart disease as a trait, like height, that exists on a continuum.[42] Both a person's DNA and a wide range of social and behavioral factors contribute to their underlying risk for heart disease. Many people assume that, like height, the distribution of heart disease *risk* in a population forms a bell curve. Polygenic embryo selection, in theory, could reduce the incidence of heart disease by changing a child's expected heart disease risk.

Unfortunately, heart disease risk is typically measured in nebulous statistical units—"standard deviations"—that are inscrutable

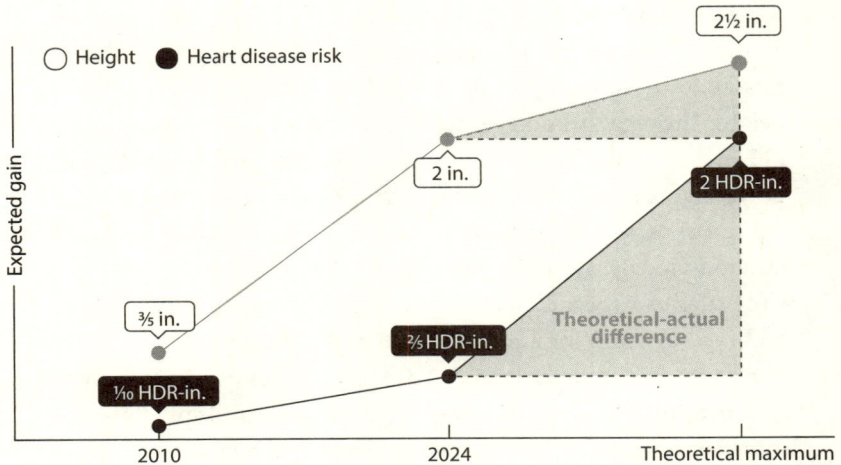

FIGURE 6. Effectiveness of polygenic embryo selection increases over time. See technical appendix for more details.

to most. As a helpful trick, you can instead think of heart disease risk on the same scale as height: inches.[43] Each person has an ingrained understanding of the distribution of human height, even if they don't realize it. Common heights for men and women span a range of 7 or 8 inches. People know that, for men's height, 6′2″ is quite tall (the ninetieth percentile) and that 5′6″ is comparatively short (the tenth percentile). Similarly, for women's height, people recognize that 5′8″ is tall (the ninetieth percentile) and that 5′1″ is relatively short (the tenth percentile). If a woman is 1 inch taller than her sister, then on average the pair are 10 height percentiles apart from one another; the same goes for a set of brothers. Thinking about heart disease risk in terms of inches—or what this chapter henceforth refers to as *HR-inches*—allows people to use an intuitive understanding of height to gauge how small or large changes in heart disease risk are.

Since heart disease is less genetically influenced than height, the impact of polygenic embryo selection on heart disease will be smaller than for height, even in the long run (figure 6). While one

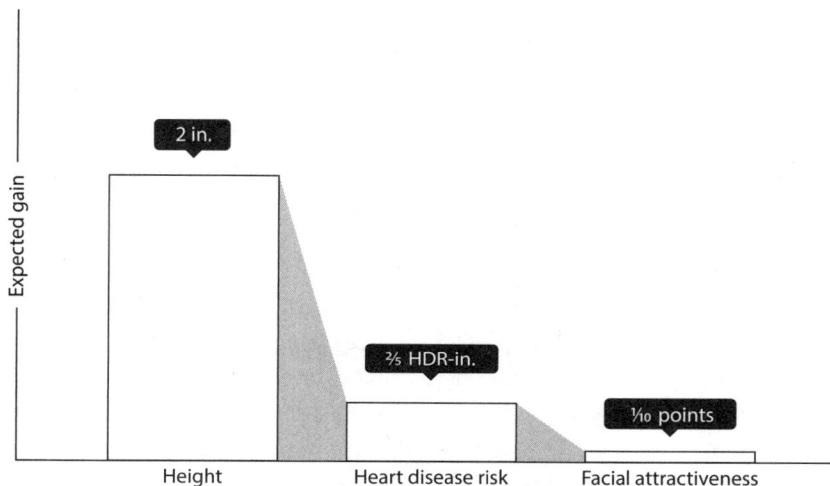

FIGURE 7. Effectiveness of polygenic embryo selection varies by trait. See technical appendix for more details.

day people *may* be able to alter expected height by 2 ½ inches, the best change in heart disease risk possible (using ten embryos) is about 2 HR-inches. Thuy and Rafal, however, were not able to use this "best possible" heart disease polygenic score. The polygenic score for heart disease that the couple used, like the polygenic scores for most traits and diseases in the world today, is nowhere near the theoretical potential. (See the technical appendix at the end of the book for more information on the construction of various estimates of expected change from polygenic embryo selection presented throughout this chapter.)

Even though heart disease is a leading cause of death in the United States, it affects less than 7% of Americans—about one in fifteen. Because of this fact, and because the diagnosis of heart disease requires the use of specialized medical equipment (such as an electrocardiogram), genomic studies of heart disease have comparatively fewer study participants. (In contrast, everyone has a height, and height is straightforward to measure.) The smaller size of genomic studies of heart disease reduces the accuracy of the resulting polygenic score. The lower accuracy of the heart disease polygenic score, in turn, severely limits its usefulness for polygenic embryo selection. If Thuy and Rafal had ten embryos to choose from (instead of four), how much of a reduction in heart disease risk should they expect to gain by selecting the embryo with the lowest polygenic score? The couple could expect that the embryo with the lowest polygenic score for heart disease would have less than a ½ HR-inch decrease in risk. According to a recent study by members of the SSGAC (Social Science Genetic Association Consortium, introduced in chapter 5) that cautions against polygenic embryo selection, this ½ HR-inch decrease corresponds to a reduction in the chances of eventually developing heart disease risk by about one percentage point.[44] At present, the polygenic scores for many traits are even less accurate than the polygenic score for heart disease; for instance, the expected gain for a trait like facial attractiveness, which has only been explored in genomic studies with a few thousand participants, is very close to zero (figure 7).

In reality, Thuy and Rafal had only four viable embryos, not ten. Thus, the expected result for their daughter Aurea is even *less* impressive: below a ⅓ HR-inch decrease in heart disease risk. Thousands of dollars and a lot of fuss—was it worth it? If Thuy and Rafal truly want to improve their daughter's heart health, would their money be better spent buying Aurea nutritious foods or swim lessons? It is unclear what polygenic-embryo-selection start-ups are promising their clients, but—with the utter lack of regulation and oversight—it is likely in a company's best financial interests to have their marketing team paint as rosy a picture as possible.

This chapter's discussion of the expected change from polygenic selection has, so far, been in reference to embryos of European ancestries.[45] For individuals hailing from other branches of the Family Tree, polygenic embryo selection is even *less* feasible. Genomic studies have largely focused on individuals of European ancestries, but most of the world's population exists elsewhere on the Family Tree. This bias poses a serious problem, which scientists call the *portability problem*. Existing polygenic scores do not accurately generalize to the majority of the global population. Large, federally funded initiatives, like the *All of Us* research program in the United States or the *Our Future Health* program in the United Kingdom, are actively working to diversify genomic data and research. Nonetheless, at present, only a fraction of the potential clinical or health benefits of polygenic scores are available to individuals of non-European ancestries.

This particular limitation of current polygenic scores makes polygenic embryo selection even less effective for non-European ancestries. For height, while current polygenic scores can produce an expected gain of 2 inches for embryos of European ancestries, the expected gain is 1¼ inches for embryos of Indigenous American ancestries, just 1 inch for embryos of East Asian ancestries, and less than ½ inch for embryos of Sub-Saharan African ancestries. Similarly, while the latest polygenic score for heart disease can decrease heart disease risk by just under ½ an HR-inch in European ancestries, it buys just a ¼ HR-inch decrease (0.9 percentage point) in Indigenous American ancestries, a ⅕ HR-inch decrease (0.8 percentage point) in East Asian ancestries, and a ⅒ HR-inch decrease (0.6 percentage point) in Sub-Saharan African ancestries (figure 8).

It is important to recall that even using the terms East Asian and Sub-Saharan African is an oversimplification (as discussed in chapter 3). Quantifying gaps in the accuracy of polygenic scores often requires categorizing individuals into discrete populations. Nonetheless, what truly produces this diminished accuracy are the messy and intricate patterns of relatedness governed by the Family Tree.[46] If most samples in existing genomic studies come from individuals of *northern* European ancestries, then this bias would decrease the accuracy of the resulting polygenic scores in individuals of *southern* European ancestries. Moreover, nowadays many people have ancestors hailing from multiple locations in the Family Tree—think of Sam, whose genome contains DNA segments from Ashkenazi, Iberian, and Indigenous American

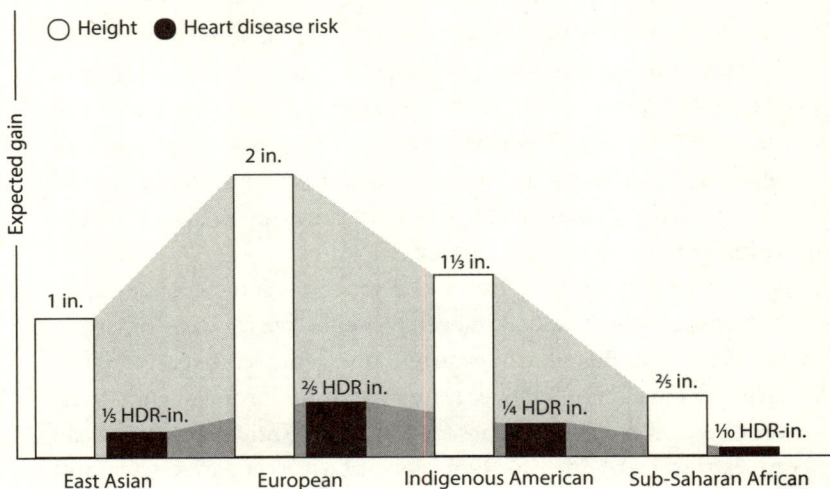

FIGURE 8. Effectiveness of polygenic embryo selection currently varies across ancestries. See technical appendix for more details.

populations, or Daphne, whose genome contains DNA segments from both West African and Eastern European populations. The usefulness of polygenic selection in embryos of diverse ancestries depends on the specific constellation of ancestries present.

It is also possible that, due to the ever-changing social and physical environment, long-run estimates of expected changes are overly optimistic. Because the effects of DNA frequently operate *through* environmental features, polygenic scores may have decreased accuracy as these environments change over time. Selecting embryos using polygenic scores essentially involves guessing which strands of DNA will be most advantageous in the coming decades, but the future is uncertain. For instance, consider an example from Sam's nongenomic research: lead exposure.[47] Lead is a powerful neurotoxin that is particularly dangerous for children, partly because children are in the midst of a sensitive period of neurodevelopment, but also due to another, less technical reason: toddlers tend to play on the ground and put objects, especially sweet-tasting ones (like chips of lead paint) into their mouths. Imagine that a person's DNA impacts the timing of developmental milestones, like how long a child spends in the crawling phase, which in turn impacts how much lead a child is likely to ingest. If lead exposure in childhood affects mental health in adulthood, any DNA effects on crawling, which in turn increase childhood lead exposure, are captured by current polygenic scores. However, the use of lead paint has been banned in the United States since 1978, and a variety of lead abatement programs have caused lead exposure to plummet. In the future, DNA variants related to crawling may no longer produce increased mental health risks, but such a process would still be captured in current polygenic scores (and therefore influence embryo selection). Only time will tell how useful current polygenic scores will be for predicting various traits and diseases in future generations.

Though proponents of polygenic embryo selection like to make the effects of DNA—as well as the task of optimizing the human genome—seem simple and straightforward, in reality, it is anything but that. When considering the scientific limitations of polygenic embryo selection, it is important to account for not just the accuracy of polygenic scores for an individual but also how that individual is embedded into a wider societal process. Social scientists have long understood that an individual's characteristics end up mattering for their life differently depending on how they fit into a larger social and economic ecosystem. Thus, applications of polygenic embryo selection that succeed in benefiting certain children may do so at the expense of others.[48]

A key example of this phenomenon involves *relative* traits. The concept of relative (or *positional*) traits dates back, at least, to Thorstein Veblen, an American economist and sociologist who wrote a treatise entitled *The Theory of the Leisure Class* in 1899.[49] If an individual benefits because their trait is greater or less than those around them, then that trait is relative. Consider a gaggle of young children playing tag at a park. Imagine that one child is doing particularly well; the group has been playing for over an hour, but she has yet to be tagged once. She runs fast for a small child, capable of sprinting about six miles per hour. What really matters for her success in tag is not how fast she can run in absolute terms, but how fast she can run *relative* to the other kids in the game. If, in a parallel universe, certain parents began selecting embryos to increase the speed of the other children playing the game, suddenly, six miles per hour may no longer be fast (in a relative sense)—at least not fast enough to avoid being tagged.

Many traits, from height to attractiveness, are relative to some extent. Often, this relativity arises when humans compete against each other. When people swipe on a dating app, what matters is how attractive a person is compared to the rest of the dating pool. Similarly, some basketball players succeed by being able reach over the heads of the members of the opposing team. Even education and work experience are, in part, relative: what connects them to labor market rewards is often where your application ranks within the broader pile. Importantly, the impacts of polygenic embryo selection on relative traits are *zero-sum* for overall well-being; any increases in well-being in certain individuals entail symmetrical losses by others. If a trait is relative, polygenic embryo selection may help certain individuals, but it is, in effect, simply shuffling around *who* benefits. This means that, for relative traits, there exists a version of the tragedy of the commons:[50] if we all reach into the cookie jar at the same time, then we'll all come back empty-handed. If basketball players all became half a foot taller, the NBA could just raise the hoops by six inches and the game would effectively be unchanged.

———

For most traits and diseases, companies offering polygenic embryo selection are currently selling consumers little more than snake oil. However, in the coming decades, the accuracy of polygenic scores will likely improve. These improvements in accuracy will mean that a wider range of characteristics will become viable targets for polygenic embryo selection, raising a host of concerns. Among them, first and foremost is the potential exacerbation and, worse still, biological reification of structural inequality that could come from unequal access to the technology.

If the United States continues on its current path, polygenic embryo selection will only be available to those with enough money to afford IVF and will—at least for a time—be most effective in individuals of European ancestries. The high costs of IVF are prohibitively expensive for working- and middle-class Americans. A single cycle of IVF costs between $15,000 and $20,000—and, at present, most couples undergoing IVF go through three or four cycles to be successful, with extra costs incurred to freeze embryos or use donor eggs.[51] (However, because these couples are typically experiencing infertility, the extent to which these figures generalize the broader American population of prospective parents is uncertain.) Private health insurance coverage of IVF is typically limited and varies across states and employers.[52] Medicaid, the public health insurance offered to low-income families in the United States, does not cover IVF at all. Polygenic embryo selection only introduces further additional costs; Genomic Prediction, for instance, charges $1,000 per embryo analyzed, and Orchid Health charges $2,500. Heliospect charges up to $50,000 to test one hundred embryos. If the status quo continues and polygenic embryo selection remains unregulated, then unequal access to the technology will cause structural inequality to grow. The racial and socioeconomic disparities of the world, both past and present, are *not* the result of systematic DNA differences across groups. If polygenic embryo selection continues to expand unchecked, then the frightening possibility exists that a new source of racial and economic structural inequality that *is*, in part, genetically produced will emerge.

As an example, consider health disparities. Because of the portability problem, polygenic embryo selection has decreased effectiveness in non-European ancestries. If, in the coming years, the use of the technology grows, those of non-European ancestries, like Pacific Islander Americans, will largely be excluded from any health benefits that embryo selection provides. Pacific Islander Americans (such as those from Guam or Samoa) are largely of Oceanian ancestries and occupy a unique portion of the Family Tree.[53] They tend to have higher rates of diabetes, high blood pressure, and heart disease than White Americans[54]—the Centers for Disease Control lists colonialism, poverty, and inadequate access to healthy foods, among other things, as key factors contributing to this disparity.[55] However, if polygenic embryo selection continues to be less effective for Pacific Islander Americans, then this community could one day have systematically higher genetic risk for chronic health conditions than White Americans with European genetic ancestry, further worsening existing health disparities between Pacific Islander Americans and White Americans.

Imagine a similar dynamic playing out in educational settings. Today, children from working-class families are nearly twice as likely to not graduate from high school compared with children from upper-class families.[56] Imagine how this disparity would grow if upper-class families (but not working-class families) were able to afford and utilize polygenic embryo selection to decrease the rate that their children suffered from learning disabilities, such as dyslexia and ADHD. Existing educational disparities between upper-class and lower-class American children would only worsen with disparate access to polygenic technologies.

Perhaps most concerning, if unequal access to embryo selection were to create class or racial disparities in genetic risk, then these differences would be passed onto future generations—potentially even compounding and accumulating over time. Richard Herrnstein and Charles Murray were dead wrong in 1994 when they wrote in *The Bell Curve* that genetic differences have naturally emerged between the American rich and poor or between White and Black Americans. (Chapters 2 and 6 discuss *The Bell Curve*.) However, if care is not exercised, genetic differences between groups of people may emerge artificially through technologies like polygenic embryo selection. Troublingly, even the inaccurate and ineffective polygenic embryo selection that is occurring in the United States right now could spur the formation of new myths about group differences in genetic risk. The outsized power of genetic myths highlights how even just the *perception* that polygenic embryo selection has produced genetic differences between groups could become a problem in and of itself. In other words, if people believe that children born via polygenic embryo selection are materially different from (or better than) children born without it, they may treat them differently—regardless of whether an actual difference exists.[57] Scholars have shown that people can use the idea of genetic difference to disguise underlying racist, classist, and sexist attitudes.[58]

Though concerns about polygenic embryo selection abound, it is important to also consider the potential upsides. Remember Sam's experiences with nerve damage and chronic pain from the previous chapter? A person's risk for chronic pain is meaningfully influenced by their DNA,[59] and Sam's mom, Nina, has also suffered from sometimes debilitating chronic pain for most of her life. Being in pain is not a relative trait; one person hurting less is not inherently accompanied by another person hurting more. In a world where polygenic scores are accurate for individuals across the entire Family Tree, polygenic embryo selection could help reduce the rate of chronic pain in future generations. In such a world, Sam would have a hard time justifying a policy that prevented parents from accessing such a technology (and would even consider using it himself). The looming challenge is figuring out for which traits and under what circumstances polygenic embryo selection is and is not permissible.

———

This book does not offer an exhaustive set of legislative solutions to the future-facing issues presented by polygenic embryo selection. However, until the scientific limitations of polygenic scores and ethical concerns about their use in reproductive decision-making are addressed, US policymakers should enact a complete moratorium on polygenic embryo selection; the United Kingdom, for instance (as well as many other European countries and Israel), has already forbid the practice.[60] In fact, the United Kingdom provides an excellent example of a viable approach for regulation. Over 30 years ago, the country established the Human Fertilisation and Embryology Authority, a regulatory body that oversees assisted reproductive technologies like IVF. The Authority regulates all fertility clinics and human embryo research in the UK, including setting standards for privacy and quality of care, granting licenses to clinics and labs, and conducting inspections. The regulatory body routinely seeks public input, helping to ensure that laws governing fertility services are influenced by those outside the medical and academic research establishment.[61]

The science of polygenic prediction is nascent and rapidly evolving. While prospective parents want the very best for their children, without the technical and scientific training that—for the most part—only researchers in genomics-related niches have on hand, it is difficult for parents to know and judge what, if anything, they are buying. Existing polygenic scores have scientific limitations that reduce their accuracy. Many polygenic scores fail to meaningfully index risk, and little is known about how useful current polygenic scores will be in different contexts and time periods. These limitations, which affect the accuracy of polygenic embryo selection, will need to be monitored on a trait-by-trait basis.

Before utilizing polygenic embryo selection, standards for scientific accuracy will need to determine which polygenic scores are sufficiently accurate and robust. That is, a change in a polygenic score ought to connote a meaningful change in the trait of interest, and the relationship between DNA and an outcome will need to hold across a range of contexts. In addition, any given polygenic trait is affected by hundreds—if not thousands—of DNA variants; these same variants are inevitably related to other traits. Researchers call this phenomenon *pleiotropy*. Selecting against one trait, such as schizophrenia, may unintentionally have consequences for the expression of another, such as creativity. For parents to make truly informed decisions, the scientific community needs a better understanding of the underlying processes that produce pleiotropic effects.

In 2024, the US Food and Drug Administration announced plans to regulate the growing prenatal testing market—pointing to one mechanism for regulating polygenic embryo selection.[62] Over the course of four years, the federal agency intends to enact measures to ensure the safety and efficacy of prenatal and other laboratory developed tests. Companies offering polygenic embryo selection may soon need to go through a review process, although the ruling is already facing legal challenges.[63] The next chapter discusses the challenges of regulating uses of polygenic scores in more detail.

Importantly, not every polygenic score that is accurate and robust *should* be offered for polygenic embryo selection. In deciding for which polygenic traits to allow selection, society will have to navigate tradeoffs between the anticipated well-being benefits of changes to a trait and the harms. For example, the United States might at some point decide that the expected reduction of Crohn's disease, heart disease, and schizophrenia in the population is significant enough to warrant the use of polygenic embryo selection. At the same time, regulators may decide that a trait like acne does not impact morbidity and mortality enough to permit targeting it. (Remember that the more traits one selects for, the less effective polygenic embryo selection is for each trait.) Relative traits like athletic performance should be prohibited altogether, as selecting for these traits would benefit some (i.e., those conceived using polygenic embryo selection) at the expense of others.

In the long term, if society approves of polygenic embryo selection for traits that meet agreed-upon scientific standards, then the practice should be made available either to everyone or to no one. If society permits IVF and polygenic embryo selection but fails to make these technologies available to anyone who wants them, then they will remain technologies of the privileged—potentially exacerbating structural inequality, producing group differences in genetic risk, and sparking new harmful narratives about genetic difference. Importantly, making polygenic embryo selection available to everyone requires polygenic scores that work for everyone, not just those whose ancestors hail from Europe.

To provide equal access to this technology, the American healthcare system would need to change significantly. Ensuring that polygenic embryo selection is truly available to all requires universal access to free, consistent, and quality reproductive care leading up to and during pregnancy.[64] This kind of structural change to the American healthcare system could have the added benefit of reducing inequalities that result from uneven access to care. Ultimately, of course, in any individual case, the final decision on whether to use polygenic embryo selection should be left to parents. The United States has a dark history of controlling reproductive decisions (remember Carrie Buck?); that road should not be revisited.

Polygenic embryo selection should not be allowed until, at the very least, these basic stipulations are met. Otherwise, unequal access to polygenic embryo selection will widen structural inequalities. Some readers may, at first blush, consider these recommendations for regulating polygenic embryo selection too strong. After all, in the patchwork US healthcare system, a wide range of technologies are accessible by some but not others—including the nerve stimulator in Sam's leg. What makes polygenic embryo selection different from these countless other medical advances? Crucially, even though it may at first seem otherwise, polygenic embryo selection is not merely a personal choice. Instead, it affects the collective "we."[65] Prospective parents are making reproductive decisions using polygenic embryo selection that will impact not only their offspring, but their offspring's offspring, and so on—genetically reshaping the human species in an enduring manner. Sam's decision to implant a device in his leg, on the other hand, will not permanently affect the biological makeup of his progeny. Moreover, DNA-based interventions like polygenic embryo selection—unlike Sam's nerve stimulator—also bring with them the additional baggage of genetic myths.

Table 7.1

A Guide for Regulating Polygenic Embryo Selection

In the short term:

- Enact a complete moratorium on polygenic embryo selection.
- Construct scientific standards for determining which polygenic scores are sufficiently accurate and robust for the selection of embryos for implantation.

In the longer term:

- To prevent the exacerbation of structural inequalities (and the creation of new genetic myths), the polygenic embryo selection moratorium should remain in place until all prospective parents have access to quality reproductive care before and during pregnancy.
- When universal access to polygenic embryo selection can be provided, slowly lift the moratorium on a trait-by-trait basis for traits that have polygenic scores that are accurate and robust in diverse ancestries.
- Selecting for relative traits, such as athleticism, should remain prohibited indefinitely (because targeting them would benefit some at the expense of others).

The practical implementation of the policy framework outlined in this chapter will not be simple, but it also is not as radical as you may think it is; in fact, a strong precedent for it already exists! As previously mentioned, the United Kingdom and numerous other countries already prohibit polygenic embryo selection (at least for now). Should a society later decide to allow polygenic embryo selection for certain traits, implementing universal coverage for the technique will be necessary. While this may be unlikely to occur in the United States any time soon, all other wealthy and industrialized countries already provide universal healthcare (which indicates that it's feasible). Though a publicly provided fertility program may seem expensive, simply reducing specific genetic variants strongly associated with deadly and debilitating diseases—such as Alzheimer's disease, lung cancer, and breast cancer—would likely allow such a program to more than pay for itself in the long run. A publicly provided fertility program would also echo other efforts by developed societies to combat declining fertility rates. Inefficiency and bureaucracy can arise from the introduction of government regulation, but society needs time to figure out how to effectively use polygenic embryo selection in a way that builds a better world for all.

Polygenic embryo selection, like IVF, may also get swept up in America's polarizing debate over abortion—further muddying the regulatory waters.[66] This is already the case with other forms of prenatal genetic testing. The Prenatally and Postnatally Diagnosed Conditions Awareness Act of 2008 requires that clinicians provide relevant information to prospective parents whose fetus appears to have a genetic condition. In reality, this information is not always provided—and in recent years, given greater restrictions on abortion at the state level, clinicians are increasingly prohibited from having these conversations with their patients.[67] The United States also will not be able to control whether and how other countries regulate the technology, meaning that if the United States bans polygenic embryo selection, prospective parents who really want it may be able to go outside the country to access it. If this practice becomes widespread, implementing a policy that gives everyone access to polygenic embryo selection would be the only way to avoid worsening structural inequalities.

The number of companies offering polygenic embryo selection is on the rise. Mainstream tech entrepreneurs are also getting in on the action: Sam Altman, the founder and CEO of OpenAI—the company responsible for creating ChatGPT—is an investor in Genomic Prediction,[68] and Anne Wojcicki, the co-founder and CEO of 23andMe, is an investor in Orchid Health. He Jiankui, the Chinese scientist who gene-edited Nana and Lulu, announced on social media that a patent he filed for polygenic embryo selection for schizophrenia had been approved. (Interestingly, the post has since been taken down.) Consumers are told to "maximize their chances of a successful pregnancy with an informed decision," and to "mitigate risks that could affect a future baby." Marketing slogans like "choice over chance," and "have healthy babies,"[69] present polygenic embryo selection as the responsible approach to reproduction. In fact, the CEO of Orchid Health, Noor Siddiqui, has suggested that polygenic embryo selection is not just an option that should be made available to all prospective parents, but instead is something that all parents *ought* to do. In an April 2025 social media post, she wrote: "No more rolling the dice. Why would you leave your kid's health up to luck?"[70] Consumer interest in the technology is also growing. The current reality is that, in many cases, prospective parents are spending thousands of dollars to receive inaccurate information.

Whether you see immeasurable promise or little upside to the proliferation of polygenic scores, regulations and safeguards are needed but virtually nonexistent. To maximize the potential of this emerging technology without exacerbating structural inequality, to enhance the safety and efficacy of society's continued transition into the postgenomic era, and to one day live in a world where everyone has equal opportunities to flourish, guardrails must regulate polygenic scores and their use in embryo selection.[71]

8

Polygenic Prediction in the Hands of Consumers and Institutions

In the Hands of Consumers

A few years after Thuy and Rafal became the first-ever couple to utilize polygenic embryo selection, Simone and Malcolm Collins decided to follow suit. The couple—alarmed by declining global birth rates and the possibility of demographic collapse—identify as "pro-natalists," positioning themselves alongside ChatGPT's Sam Altman and the controversial South African businessman turned political financier Elon Musk.[1] (Experts currently predict that the global population will peak in 50 years at around 10 billion people, but exactly how the population will change after that is less well understood.)[2] Simone and Malcolm believe many existing policies unfairly burden parents and daycare providers, thereby lowering fertility rates. The pair intends to do their part by having seven children; at the time of writing this book, they have four.

The couple already had two sons by the time they heard about polygenic embryo selection: Octavian and Torsten (or Toastie, for short). At that point, Simone had also undergone five rounds of IVF, costing nearly $100,000. Nonetheless, the couple decided to pursue a sixth round to utilize the new technology. Like Rafal, the two talk about polygenic embryo selection with enthusiasm: "No one else is willing to talk about [polygenic embryo selection]," Simone shared on the podcast she co-produces with her husband. "We wanted to shout from the hilltops."[3] To Simone and Malcolm, polygenic embryo selection is a win-win; it provides their offspring "the best possible roll of the dice" while also mitigating against the "genetic death spiral" they believe humanity is hurtling toward. In an online post, Malcolm attributes this impending death spiral to low fertility rates among "high-productivity groups" who have "the capacity for self-control." In contrast, he explains, those who are "less economically and intellectually productive" are reproducing at higher rates, thereby outpacing "high-earning technophilic groups."[4] While Malcolm doesn't label any specific groups as more or less productive, at a high level his dysgenic concerns resemble a key fear presented in the great replacement theory (introduced in chapter 4): that higher birth rates among members in one group will ultimately lead to the extinction of another.

Just as Genomic Prediction had done for Thuy and Rafal, the company provided Simone and Malcolm with a genetic report on each of their viable embryos. However, Simone and Malcolm felt that—before making a final decision about which embryo to implant—there was still more to learn. So, the couple downloaded each embryo's DNA data from Genomic Prediction and uploaded it to the website of a direct-to-consumer genetic testing company called SelfDecode.

Many of the most popular direct-to-consumer (or DTC) genetic testing companies, such as Ancestry.com and 23andMe, ask customers to send in raw saliva samples for genomic analysis. A growing set of companies—called *third-party* DTC genetic testing companies—do not conduct genetic sequencing (or genotyping) themselves. Instead, they ask consumers to upload a file containing their DNA data, which can be downloaded from the websites of other genetic testing companies. SelfDecode—the Florida-based company used by Simone and Malcolm—gives consumers either option: consumers can spit in a tube or upload their data file and

Simone and Malcolm Collins

gain access to hundreds of genetic risk reports ranging from anxiety to intelligence to athletic performance.

Simone and Malcolm chose to analyze their embryos' DNA for an array of mental health conditions, including schizophrenia and chronically low mood.[5] To their delight, Embryo #3—the same one selected by Genomic Prediction's polygenic health report—had the most favorable mental health results on Self Decode's polygenic assessment, making their choice clear. Less than a year later, their baby girl was born. They named her Titan Invictus Collins, eschewing traditionally feminine names that they feared others would take less seriously.[6]

In turning to SelfDecode, Simone and Malcolm used DTC genetic testing to implement a do-it-yourself version of polygenic embryo selection. Though these platforms are built for consumers looking to learn about their own DNA, the couple instead uploaded the genomes of their embryos—and it wouldn't be their last time. A couple of years after Titan was born, Simone and Malcolm used polygenic embryo selection to have their fourth baby: a girl named Industry Americus. When asked for which traits they selected in their second child, Malcolm said, "Obviously, we looked at IQ." For the founders of the Pronatalist Foundation, DIY polygenic embryo selection could amplify "genetic IQ advantage . . . much faster" than traditional reproduction; at a personal level, they also see it as a way to increase the likelihood that their genetic line will "be among those humans who colonize the stars."[7]

DTC genetic testing works just as the name suggests: rather than consulting a medical professional, you instead order a genetic test directly from a company. The very first DTC genetic testing companies launched in the late 1990s, right around the time that *Gattaca* was released. Suddenly, people had access to a new and exciting resource—information about their own DNA (and, to some extent, the DNA of their relatives)—all from the comfort of home.

Today, many companies are capitalizing on the growing availability of genomic data, and DTC genetic tests are more affordable than ever. Fifteen years ago—around when New York Fashion Week hosted a "spit party" so that models and celebrities could get in on the personal genomics movement—the average test cost about $1,000.[8] Ten years ago, the price had already dropped to roughly $100. Now, for as little as $10, DTC testing companies offer to transform a person's DNA into a commodity and a social activity.[9] Friends gift each other DNA kits as birthday presents, and universities offer courses where students are given the ability to study their own genomic data.[10]

It seems possible to find a DTC genetic test for just about anything. Alongside genetic tests for medical conditions like prostate cancer and Parkinson's disease, many DTC companies offer absurd information like whether your ancestors were farmers or hunter-gatherers or what someone's genetic breast size is.[11] Some dating companies even offer to analyze the "love genes hidden in your DNA" to provide matches whose genome will complement your personality.[12] The companies selling this science as a service tend to cleverly misconstrue just how these tests work—and how accurate they actually are.

While DTC genetic tests are currently available for a wide range of traits, many provide information that is neither scientifically valid nor practically meaningful. The scientific limitations of existing polygenic scores that reduce their effectiveness for embryo selection also cause many current DTC tests to be highly inaccurate. While clinical traits like heart disease garner large amounts of funding and scientific attention, genomic studies of nonclinical traits often have smaller sample sizes and even less rigorous results. DTC testing companies will offer consumers these crummy tests, nonetheless. For instance, one company currently offers consumers information on their DNA variants for facial attractiveness, despite the fact that the referenced genomic study has a tiny sample and therefore very little scientific information.[13] In fact, the study's results would produce a polygenic score so inaccurate that a person with a low score (tenth percentile) and a person with a high score (ninetieth percentile) would likely only have a miniscule difference in how others actually rate their attractiveness: about *a tenth* of one point on the classic 1 to 10 "hot or not" scale! Moreover, many companies provide consumers results for just a single or a few DNA variants out of the thousands possible, even though the impact of each variant is tiny and the results are effectively meaningless.[14]

Despite concerns about the validity of many DTC genetic tests, the industry's offerings continue to proliferate while much-needed oversight and regulations lag far behind. A couple of decades ago, researchers and industry insiders would have been shocked to learn that the DTC genetic testing landscape would transform into a virtually lawless frontier. In the industry's early days, government agencies seemed committed to regulation and oversight. In 2005, just a year after the founding of the first DTC company in the United States, a congressional watchdog called the US Government Accountability Office began to scrutinize the emerging industry. In 2010, the US Food and Drug Administration (FDA) informed five DTC testing companies, including 23andMe, that their tests needed to be validated and then approved by the agency. 23andMe, for instance, had been selling genetic tests for health-related conditions like Alzheimer's disease, breast cancer, and bipolar disorder.[15] By advertising that their tests provided consumers with important information about their health, DTC companies were technically selling medical devices—placing them under the jurisdiction of the FDA. Years passed with little indication that these companies planned to meaningfully heed the FDA's warning. In the summer of 2013, 23andMe launched a new promotional campaign that advertised its personal genetic services for use in "the cure, mitigation, treatment, or prevention of disease,"[16] blatantly disregarding the FDA's authority. The agency, growing increasingly concerned that the use of unvalidated genetic tests may lead consumers to make medical decisions using faulty information, had finally had enough.

In 2013, the FDA sent Anne Wojcicki, the CEO of 23andMe, a cease-and-desist letter. The letter ordered 23andMe to immediately discontinue its health-related genetic tests for hundreds of conditions and diseases. Just like a company that produces medical thermometers or surgical devices, 23andMe would need to obtain FDA approval for its products before they could once again sell them to consumers.[17]

Two years later, in 2015, 23andMe received approval to launch a revamped version of its platform—allowing consumers to undergo DTC carrier screening for rare monogenic conditions like Bloom syndrome, Tay-Sachs, and cystic fibrosis. Then, in 2017, the FDA approved 23andMe's release of ten additional tests, including ones for Alzheimer's disease and Celiac disease. Soon after, the company received approval for DTC genetic tests for breast cancer, colon cancer, and prostate cancer. However, in the time since its dramatic face-off with 23andMe, the FDA appears to have largely given up on regulating the DTC industry. To date, 23andMe, which declared bankruptcy in 2025, is the only company that has received FDA approval to offer DTC genetic tests for medical conditions. You wouldn't have to look far—only around—to find companies selling genetic tests that have not gone through any kind of regulatory approval process.

In the early days of DTC genetic testing, companies sold tests for just a single (e.g., APOE4 for Alzheimer's) or a few (e.g., BRCA 1 and 2 for breast cancer) DNA variants. Today, polygenic scores are rapidly becoming the key tool used by DTC companies to provide information on complex traits. By adding polygenic scores to their menu, companies can greatly increase the number of traits for which tests can be offered. 23andMe, for example, offers polygenic scores for anxiety, high blood pressure, seasonal allergies, and many other conditions. Strikingly, none of the polygenic scores offered by DTC companies—not even 23andMe—have gone through the FDA approval process. Why hasn't the FDA asserted any kind of regulatory authority over DTC polygenic scores?[18]

One reason is that the FDA only reviews genetic tests for *medical* traits and diseases, which means there is currently no oversight of DTC genetic tests for most traits, including characteristics like intelligence, fear of heights, or musical ability. Even in the specific case of polygenic scores for diseases, like type 2 diabetes, companies have found clever ways to evade regulatory oversight. One strategy is to blur the line regarding what exactly counts as medical, per se. Companies will market polygenic tests as wellness rather than medical products—a genetic report for "heart health" instead of heart disease or for "brain health" instead of Alzheimer's.[19] By using the term "health" instead of "disease," such companies seek to place genetic tests for single variants and polygenic scores into the category of unregulated wellness products, akin to vitamins and herbal supplements. Unless a wellness product has been proven to cause adverse side effects, the FDA generally does not review it.

Another way that companies circumvent regulation is to use polygenic scores instead of genetic tests for only a single or a few DNA variants. The general sentiment is that, because polygenic scores are calculated using algorithms, they constitute software. The FDA generally only regulates software that is intended to diagnose, prevent, monitor, treat, or alleviate disease. To dodge the FDA, a company simply needs to flash a disclaimer that its products are for educational and informational purposes only and should not be used for any diagnostic or curative purposes. The FDA has also implied that if a software simply matches findings from scientific studies with patient-specific medical information, then it will not police it.[20] Many DTC genetic testing companies are doing precisely this: finding a genomic study—for instance, one on "economic and political preferences"[21]—and then delivering consumers a genetic report on political affiliation using the resulting polygenic score.[22]

Multiple regulatory loopholes made it possible for Simone and Malcolm to embark on DIY polygenic embryo selection. First, when they uploaded their embryos' DNA files to SelfDecode, there was no need for any additional laboratory-based testing. In the eyes of the law, the company did not employ any kind of medical device (e.g., a SNP chip) to deliver results to Simone and Malcolm. Instead, it used software. Some experts speculate that even when companies conduct in-house DNA sequencing / genotyping, polygenic scores are still technically software because they are generated using a computational procedure.[23] In practice, the distinction between a polygenic score as software and a polygenic score as a hardware medical device is meaningless; polygenic scores are ultimately calculated the same way regardless of whether a company processed the biological samples itself. In addition, SelfDecode issues a disclaimer flagging that its reports are educational and recommending that customers consult their healthcare providers, stating: "our services are not to replace the relationship with a licensed doctor or regular medical screenings."[24]

Patchy, inconsistent, or nonexistent regulation is a problem for a few reasons. The first is that DTC genetic tests often provide inaccurate information,[25] and many consumers may still make healthcare decisions using their results. For instance, using a DTC genetic test for sensitivity to anticoagulant medications, a person might decide to abandon certain drug therapies. Or, after receiving the results of a genetic test for breast cancer risk, a person may decide to undergo a double mastectomy. (These two examples were included in the FDA's 2013 letter to 23andMe.[26]) Researchers have found that consumers tend to respond to genetic tests about their health in a variety of relatively low-stakes ways, like changing their diet or seeking guidance from genetic counselors.[27] In some cases, though, individuals have taken more drastic measures. For instance, a twenty-two-year-old man took medical leave from his PhD program and stopped rigorous exercise after his DTC data suggested he carried a genetic variant associated with heart failure; confirmatory testing conducted by a medical specialist later determined that the result was a false positive—much to his relief.[28] The powerful grip that DNA has on the popular imagination serves to strengthen the impact that DTC genetic tests have on those who take them. Researchers have shown negative psychosocial effects to receiving information about genetic propensity for traits like educational attainment or depression.[29] In addition, as more and more people access DTC genetic testing and bring their results to their doctors, some medical professionals worry that it could lead to unnecessary follow-up tests and even harmful medical procedures—potentially adding undue strain to the healthcare system.[30]

A second problem surrounding the lack of DTC regulation is the proliferation of genetic myths, which threatens to validate or exacerbate structural inequality. While inaccurate genetic tests may not provide consumers with meaningful information about their risk or predispositions, they can still lead people to emphasize the effects of DNA. In doing so, they may add fuel to existing beliefs in genetic myths. In some cases, companies explicitly reinforce such myths via their marketing of DTC tests. For instance, one company offers a genomic test for political affiliation and advertises that learning that a person's politics are "determined by genetics can make it easier to understand their point of view" and "may make us stop trying to change each other's perspectives so aggressively."[31] While this promotional strategy may be intended to encourage tolerance, it espouses the Destiny Myth, leading consumers to believe that, if political views are influenced by DNA, then they are inevitable and unchangeable. Ironically, while genetic myths regarding social and behavioral traits are historically the most pernicious, these are the very traits that fall outside the FDA's purview because they are typically not considered medical. Loopholes like these must be closed, but doing so is easier said than done.

To start, commonsense regulations that cover the *entire* DTC genetic testing industry must be put in place. If DTC companies continue to elude regulation, there is little hope that consumers will be able to accurately discern *which* tests have *how much* valid information. Importantly, any regulatory distinction between whether tests are medical versus nonmedical, or software versus hardware, must be reconsidered. Consumers use DTC genetic tests to make decisions about their own health and well-being regardless of whether those tests are technically medical or whether the company that provides the test results also conducts the sequencing /genotyping. Furthermore, DTCs of medical and non-medical traits alike have the potential to activate or reinforce genetic myths. Scientific standards and approvals must be created to ensure snake-oil genetic tests never make it to market.

There are indications that such changes are beginning to occur. As mentioned in chapter 7, in 2024, the FDA announced plans to begin regulating laboratory developed tests that are intended for clinical use (like non-invasive prenatal genetic tests). This shift was spurred, in part, by the recognition that laboratory-developed tools are increasingly being used without any measures to ensure their safety and efficacy. With these changes, which will be rolled out over four years, the FDA now explicitly categorizes in vitro diagnostic products like laboratory developed tests as devices.

US–based DTC companies that perform in-house sequencing or genotyping will need to comply with FDA requirements for such devices, such as premarket review and labeling requirements. However, because the FDA specifically declared its authority to oversee *laboratory-developed* tests, it remains unclear whether DTC companies that do not collect and process genetic data themselves must comply with these regulatory changes. Mounting legal challenges also claim that the FDA does not have the authority to regulate the professional testing services provided by laboratories;[32] it remains to be seen whether this ruling will stand.

It is unlikely that the FDA's regulatory updates (should they withstand legal challenges) will prevent the sort of do-it-yourself polygenic embryo selection that the Collinses utilized. DTC companies currently have no way of tracking whether the information they provide is being used by consumers for prenatal testing. The door to DIY polygenic embryo selection needs to be firmly shut. Getting polygenic embryo selection right for society means that preventative measures must keep prospective parents from downloading their embryos' DNA data files and uploading them to DTC genetic testing companies. Even if society carefully restricts polygenic embryo selection, an unrestricted DTC genetic testing industry will allow people to covertly select for traits not offered by polygenic embryo selection companies (just as Simone and Malcolm did).

When utilizing polygenic scores for embryo selection, regulations should ban healthcare providers and scientific laboratories from providing the raw DNA data of the tested embryos to prospective parents and should instead provide only reports on genetic risk. Once an embryo is selected and implanted, though, the risks of allowing parents to access the implanted embryo's DNA data seem to decrease significantly.

There are other gaps and loopholes of which policymakers will need to be aware. As with polygenic embryo selection, consumers will likely continue to access DTC genetic testing services from non-US countries where regulations vary substantially.[33] A thorough regulatory response to DTC genetic testing would restrict the websites of non-US DTC genetic testing companies whose services have not passed US regulatory guidelines. The nation took a similar approach in 2006 to prohibit unlawful internet gambling sites.[34]

Nevertheless, if a DTC genetic test is scientifically accurate and its results appropriately communicated, it is difficult to justify not allowing individuals to access the information stored in their own genome—especially if public input on the matter reveals a strong preference for knowing. However, DTC companies must work to ensure that their products do not disseminate genetic myths. To support accurate interpretation of difficult results, DTC companies should offer consumers consultation services. Accompanying educational literature must accurately document the limitations of the tests and for whom the results may be inaccurate. 23andMe, for instance, offers in-depth documentation and referrals to consultations (though consultations may come with additional costs).[35]

Finally, while individuals can choose whether they participate in these various forms of DTC testing, their relatives do not. An individual's decision to provide their DNA necessarily implicates their biological relatives; this is how the Golden State Killer—a prolific serial rapist and murderer who terrorized California in the 1970s and '80s—was caught.[36] Although there are benefits to using consumer DNA, such as tracking down dangerous criminals, there are also questions of agency and privacy. Moreover, DTC testing does not just bring into question the privacy of biological relatives: in 2023, 23andMe experienced a colossal security breach of almost 7 million of its customers' DNA data. Then, in 2025, the company declared bankruptcy after struggling for years with declining sales (which, in part, results from the fact that it sells a test that consumers only need to purchase once in their lifetime). As of the writing of this book, 23andMe—and its massive database of genomic data—is up for sale to the highest bidder.[37] Some consumers are concerned about the privacy implications of the sale of their 23andMe data. In June 2025, a number of states and the

Table 8.1

A Guide for Regulating Direct-to-Consumer Genetic Testing

- Develop scientific standards for determining which polygenic scores are sufficiently accurate for provision to consumers.
- Produce communication guidelines regarding how companies can best describe the limitations of DTC tests and avoid the proliferation of genetic myths.
- Close loopholes that allow polygenic scores to remain unregulated by abolishing the distinction between (i) medical vs. nonmedical; and (ii) software vs. nonsoftware in the case of DTC genetic tests.
- Restrict access to websites of foreign DTC testing companies who do not meet US regulatory guidelines.
- Prevent backdoor polygenic embryo selection by prohibiting prospective parents from accessing raw embryo DNA data prior to implantation.

District of Columbia sued 23andMe—opposing the sale of consumer data without their consent.[38]

DTC genetic testing has been around for decades. In that time, little regulation of the market has gone into effect. Ample regulatory loopholes exist for companies, especially those using polygenic scores, to evade even the limited oversight that does exist. In the meantime, people are utilizing DTC services that frequently provide inaccurate or misleading information. If you have done DTC testing, take your results with a hefty ladle of salt. If you have not, a word of caution: there are privacy risks associated with giving your DNA to an industry concerned primarily with making a financial profit.

In the Hands of Institutions

While consumers learn to navigate the fast-moving and often bewildering landscape of DTC genetic testing, institutions are beginning to grapple with whether and how to utilize new genomic

tools. Perhaps the highest profile institutional effort to apply polygenic scores is in *precision medicine*. Doctors at many hospitals and
clinics (and at the US Centers for Disease Control) believe that
incorporating polygenic scores into existing health screening programs could improve the early diagnosis and treatment of disease.[39] If effective, polygenic-informed screenings could reduce
social inequality (broadly defined) by targeting interventions to
those with the most need. Because any benefits of incorporating
polygenic scores into screenings hinge on their ability to identify
previously hidden risk factors, though, utilizing such scores offers
little in terms of reducing structural inequality (which usually exists across visible and stigmatized characteristics).

Polygenic scores are starting to be used to screen for heart disease and various cancers in adults and type 1 diabetes in new-
borns.[40] An open question remains as to whether this genomic
technology can meaningfully improve clinical care (and for which
diseases). For example, if a polygenic score for a given disease is
inaccurate, or if existing clinical measures sufficiently capture
underlying disease risk, then using polygenic scores could increase
the cost and complexity of screenings without actually improving
clinical care.[41]

Although polygenic-informed screenings are currently being
rolled out in medical contexts, they do not, in theory, need to be
confined to them. When Daphne was a PhD student, she frequently browsed the shelves of the Cambridge University library—
or, as students and faculty called it, "the UL." Housed in a mam-
moth brick building, the UL contains over 9 million texts. One day,
while perusing the poorly lit stacks, Daphne stumbled upon a
book that transformed her entire research and career trajectory:
*G Is for Genes: The Impact of Genetics on Education and Achieve-
ment*.[42] The book's authors mount a controversial argument, calling on educators and policymakers to consider the promise of
"*precision education*"—in which every student receives an education plan specifically tailored to them using their DNA. The idea
that genomic data and polygenic scores could inform how humans

design and operate educational systems may seem as unlikely now as genetic embryo selection did to *Gattaca* viewers in the late '90s.[43] However, in the 2010s, Yale researchers teamed up with local public schools to start building a "genetic screener for dyslexia."[44] Early identification of dyslexia is key, but diagnoses typically occur only after a child starts falling behind in reading—too late for optimal intervention. Such a genetic screener would likely garner wide appeal, as a majority of American parents report interest in using DNA to test for their children's learning disabilities.[45]

It may be only a matter of time before well-resourced parents decide to use DTC genetic testing to ask schools for additional resources for their children: extra time on tests for Cindy who has a high polygenic score for ADHD, or anti-anxiety medication for Oliver who has a high polygenic score for neuroticism. As Daphne encountered during her PhD research, some schools already only admit so-called gifted students; it is not so far-fetched to imagine that such schools may one day use polygenic scores when making admissions decisions or allocating educational resources. In fact, in 2019, a prominent British-American researcher (one of the authors of *G Is for Genes*) received preliminary approval to implement a controversial new program with the goal of identifying impoverished children with high polygenic scores for educational traits and investing extra resources in their development.[46]

Even if schools decide against implementing polygenic-informed screenings, ripples from the DTC industry may force such institutions to begin interacting with polygenic scores. Professional societies like the American Society of Human Genetics have cautioned against the use of pediatric genetic testing for nonmedical purposes or without a demonstrated medical need.[47] Nevertheless, in China, some parents have used DTC genetic "talent" tests to try to identify the optimal educational niches for their children.[48] In the United States, some parents have purchased DTC tests to help decide which sports to enroll their children in.[49] Certain DTC companies are even marketing offers to "unlock your child's success" and "discover your child's hidden genetic talents."[50]

It's also important to remember that schools are, in part, beholden to the desires of parents. When parents are unhappy with the options available to them, they can decide to leave the traditional education system altogether and strike out on their own. For instance, Simone and Malcolm Collins—the couple that utilized DIY polygenic embryo selection—have taken that same DIY approach to their children's education. They feel that America's traditional schooling system is "optimized to turn humans into replaceable parts"[51] and not to "foster individual strengths, independent thought, emotional control, and internal motivation."[52] Wanting more for their children, Simone and Malcolm founded their own school in 2022: the Collins Institute for the Gifted. From leveraging new genomic technologies to demanding more from their child's school, some parents will go to great lengths to try to optimize the health and success of their offspring.

Whether in the doctor's office, the classroom, or another setting seemingly unrelated to genetics, polygenic-informed screenings— even ones that intend to offer extra resources to those who need them most—present potential pitfalls. Depending on the trait, having a high or low polygenic score could be stigmatizing in and of itself, regardless of whether the trait in question ever manifests. Unless it is emphasized that polygenic scores are probabilistic, not deterministic, implementing polygenic-informed screenings could stoke the fires of the Destiny Myth. Having a high polygenic score for heart disease does not guarantee that a person will suffer a heart attack, but everyone that has the specific DNA repeat that causes Huntington's disease will eventually develop the disorder; in a world with easy access to polygenic scores, everyone will need to learn this important distinction. If they don't, how might polygenic scores shape people's beliefs about themselves and others? As Rosenthal and Jacobson showed in chapter 1, a teacher's beliefs about her students affect their educational trajectories, and as

Brianna illustrated in chapter 4, some children internalize the belief that they are less capable from a young age. Ideas have a tangible impact on how people live their lives and treat each other.

If used inappropriately, polygenic scores could also distract from key social and environmental risk factors—like poverty. (This concern also features in the debate over DNA and social inequality that was covered in chapter 6.) Moreover, any benefits of improved screening will only emerge for those with the means to access the ensuing intervention, unless the relevant interventions are made widely accessible. Finally, as the previous chapter pointed out, the majority of the benefits of polygenic-informed screenings will also be, at least in the short term, concentrated among individuals of European ancestries. Given these risks, how can polygenic-informed screenings be implemented responsibly?

If institutions plan to use polygenic-informed screenings, they must do so carefully. To that end, a polygenic score should never serve as the only piece of information used in a screening. They should always be combined with other relevant risk factors—always polygenic-*informed* screening, never polygenic-*only* screening. Using multiple sources of information will both increase accuracy and make screenings less likely to reinforce genetic myths. In addition, the relative accuracy of any risk factor included in a screening should be validated to ensure that weak predictors—like an inaccurate polygenic score—are not prioritized over robust ones. Polygenic-informed screenings should also only be utilized when there is an effective intervention; they should not be implemented for informational purposes only.[53] Finally, special care must be taken to ensure that the decreased accuracy of polygenic scores for individuals of non-European ancestries does not induce disparities in resource allocation across key structural categories. That is, a school's inability to use polygenic scores to accurately assess one group's risk for dyslexia, for example, should not result in fewer reading coaches for students in that group (compared to other groups).

Importantly, polygenic-informed screenings must be considered a form of *black box prediction*. Black box prediction, a term from the field of machine learning and artificial intelligence, describes technologies that offer prediction without providing a clear understanding as to why. Such opaque technologies are rapidly evolving and proliferating: for instance in the content algorithms used by streaming services to recommend TV shows and the facial recognition software that allows people to unlock their phones at a glance. The myriad DNA variants captured by a polygenic score operate through labyrinthine processes that are largely unknown, making polygenic scores subject to the same critiques and limitations as other black box predictors. Although this book does not delve into the broader debate regarding black box prediction, it is worth noting that the suitability of black box prediction depends on the context. Black box prediction may be harmless enough for Netflix's content algorithms but is less benign for making bail decisions or deciding to whom to prescribe a certain antidepressant.

Polygenic-informed screenings are frequently presented as a method for identifying and helping high-risk individuals. However, the technology could just as easily pile on privilege for the most advantaged, thereby exacerbating social inequalities. Without proper consideration and regulation, DNA could come to stratify and segregate society as race and class already do. A person's polygenic score would not just be (scientifically speaking) *higher* or *lower*—but instead (normatively speaking) *good* or *bad*, *superior* or *inferior*. The world could come to resemble a dystopia not so unlike *Gattaca*, where DNA is analyzed and used to determine possible educational or career paths and where those with the "wrong" genes face stigmatization and discrimination. If banks start using polygenic scores for risk-taking behavior to evaluate loan applicants,[54] or if basketball scouts begin ranking youth basketball players using their polygenic scores for height, then polygenic score information could become a structural category. Importantly, this social shift could occur regardless of whether the polygenic scores used are scientifically accurate.

In the movie *Gattaca*, genetic discrimination is technically illegal. Still, institutions find covert ways to get around the law and obtain a person's DNA: skin cells are collected after a handshake, and urine samples from mandatory drug tests are repurposed. In the United States today, legal protections against genetic discrimination are so limited that there would be little need to resort to subterfuge. There is just one relevant piece of federal legislation: the Genetic Information Nondiscrimination Act, or GINA for short. GINA was passed in 2008 and prohibits employers and health insurers from discriminating against individuals using genetic information.[55] However, GINA does not protect against genetic discrimination in most settings, including life insurance, schooling, housing, financial lending, and the military. As the applications of genomic tools like polygenic scores grow (and, in turn, occurrences of genetic discrimination multiply),[56] expanding genetic discrimination laws in the United States is becoming increasingly important. On this front, two states on opposite poles of the divided political landscape are leading the way. In 2011, California amended its existing anti-discrimination laws to prohibit genetic discrimination in a variety of domains, including mortgage lending, education, and housing. In 2020, Florida enacted legislation prohibiting life- and long-term care insurance companies from discriminating based on genetic information.[57]

A key challenge for policymakers will be striking the difficult balance between preventing the use of polygenic scores to genetically discriminate while still permitting certain uses of polygenic scores to screen for individuals who may benefit from specific interventions or from receiving extra resources. It will take time to develop the exact legal and regulatory framework required to achieve such a balance—but precedents do exist. For instance, the Americans with Disabilities Act prohibits discrimination against people with disabilities while simultaneously guaranteeing certain accommodations—be it ramps and elevators or a reading coach—to those who need them.[58] Similarly, the Fourteenth Amendment and Civil Rights Act protect against racial discrimination (for instance, a university refusing to admit Hispanic students) while still allowing many forms of affirmative action (race-conscious efforts

Table 8.2

A Guide for Regulating Polygenic-Informed Screenings

- Always combine polygenic scores with other relevant risk factors for the purpose of screening.
- When useful, utilize polygenic score–informed screening to grant additional assistance to those with elevated risk of a negative outcome (e.g., targeting interventions) but not to confer advantages to those with elevated risk for a positive outcome (e.g., genetic discrimination).
  - Only implement a polygenic score–informed screening when there is an effective potential intervention.
  - Ensure that any differences in the accuracy of polygenic scores as a function of ancestry do not induce resource allocation imbalances across key structural categories.
- Expand federal and state genetic nondiscrimination laws to include settings like schools, financial lending, and life insurance.
- Recognize that polygenic scores are a form of *black box prediction.*

to increase the number of members of a group that have been historically underrepresented in a given industry).[59] As a rule of thumb, polygenic scores should only be leveraged to grant additional help to those who have an increased chance of a negative outcome and should never be used to confer an advantage to the already advantaged.

———

While the regulation of polygenic embryo selection, DTC genetic testing, and polygenic-informed screenings may be difficult, it is certainly not impossible. (Think back to the UK's Human Fertilisation and Embryology Authority from the last chapter.) The United States already has statutory bodies like the Food and Drug Administration, the Centers for Medicare and Medicaid Services, and the Federal Trade Commission, whose functions center on considering evidence and establishing rules that protect patients and consumers. The Centers for Medicare and Medicaid Services, for instance, is responsible for the implementation of federal regulations focused on the analytical validity of diagnostic tests and laboratory processes. While DTC companies like Ancestry.com conduct testing in laboratories certified by these regulations, companies based outside of the United States and those that are not handling biological samples do not face such oversight, even if the product they deliver to consumers—polygenic score results—is ultimately the same. Furthermore, the Federal Trade Commission exists to enforce consumer protection laws and prevent fraudulent, deceptive, and unfair business practices. It is certainly within the agency's scope to prevent, monitor, and take any necessary legal action against DTC companies that advertise DNA as prophecy, oversell their products, or overstate their results. Although the FTC does provide some recommendations to companies selling genetic testing kits (e.g., "tell the truth about what your genetic testing kit can do"), there is little formal oversight.[60]

Regulatory oversight is vitally important, but additional steps will help people appropriately understand the risks, benefits, and limitations of polygenic scores. It is essential to alter how genetics is taught in K-12 education. Gregor Mendel and his peapods, which many American children learn about in school, offer little reassurance when it comes to understanding the intricacies of complex traits and polygenic scores. This educational approach means that most people leave school with a limited understanding of genes and how they work.[61] It is increasingly important that everyday Americans can discern the difference between using a genetic test to select an embryo without sickle cell disease and selecting an embryo because it has the lowest polygenic score for Type 2 diabetes—and that polygenic scores only provide probabilistic guesses, never guarantees. In short, genetic literacy on the part of consumers and policymakers, as well as responsible advertising from companies providing these services, are important for supporting informed choices regarding genetic information—regardless of whether that information is making its way to consumers via polygenic embryo selection, DTC testing, or polygenic-informed screenings. To support genetic literacy, children deserve and require a more thorough and accurate genetics education.

A nuanced understanding of the complexities of human genetics would not only equip people with the background and context necessary to understand the limitations of polygenic scores and their applications; research suggests that it could also help dispel genetic myths.[62] For instance, studies have shown that reading materials that describe DNA as a "blueprint" or that include "gene for" language increase belief in the Destiny Myth.[63] In contrast, people who learn about patterns of human genetic variation and the polygenic nature of most human traits are less likely to subscribe to the Destiny and Race Myths; the more a person understands the inherent flaws in genetic myths, the less likely they are to believe in them.[64] Preliminary attempts to revamp K-12 genetics education are underway, but additional efforts are needed in order to secure the responsible proliferation and interpretation of polygenic scores in the world. As these genetic technologies become increasingly available, increased genetic literacy is key to ensuring every person can make an informed decision about whether and how best to use these new resources.

9

The Future

While this book has surfaced a number of enduring disagreements, the writing process has also revealed a shared core belief: in order to fully understand the role that genes play in the world, people must take seriously *both* acids that humans inherit: the biological processes of DNA and their outsized role in the social imagination.

Just over two decades ago, the completion of the Human Genome Project forever changed the field of human genetics. The As, Cs, Ts, and Gs of millions of genomes have now been assayed and stored in large databases. In turn, this information has advanced our collective understanding of the complex ways that the DNA a person inherits affects their life. Layered on top of biological variation in people's DNA profiles are myths that societies and their members create, narrate, and pass down about genes and how they affect people. Historically, genetic myths supported efforts to involuntarily sterilize a teenager and contributed to the late-night arrest of an interracial couple sleeping in their bed. More recently, they sat in on a controversial conversation between an alt-right podcaster and a physics professor at a well-respected university. When White supremacists chugged large amounts of milk as a supposed demonstration of their racial and genetic superiority, genetic myths were present. And more tragically, such myths played a part in an online justification of the murder of ten Black Americans in a grocery store.

Humans inherit DNA, the literal acid in cells, as well as genetic myths, a conceptual acid in society. The rapid proliferation of polygenic scores, a new tool for making predictions ranging from how far someone goes in school to their likelihood of developing heart disease, carries both promise and peril. What's clear is that the genie is out of the bottle—taking shape in polygenic embryo selection, direct-to-consumer genetic testing, and polygenic informed screenings. How can and should society respond to this changing technological landscape?

––––––

Genes are not destiny, and race is not biological. Part 1 of this book debunked two long-standing genetic myths, which have been socially inherited from a dark past riddled with eugenics. Chapter 2 focused on the Destiny Myth, which falsely claims that the effects of DNA are immutable and inevitable. Prior to the twenty-first century, while scientists knew that genetic factors played a role in shaping a wide range of individual characteristics, there was no way to identify which specific genes mattered. Over the past two decades, a deluge of new genomic data has led to rapid scientific advances in the ability to connect specific DNA variants to a range of valued outcomes. Still, these genomic discoveries remain largely a black box; the precise pathways connecting DNA to eventual life outcomes are largely unknown and may remain so. Because humans have the bad habit of thinking that the effects of DNA are simple and straightforward, these discoveries might usher in feelings of fatalism and hopelessness.

By understanding and utilizing the style of counterfactual thinking that undergirds recent scientific advances, one can plainly see that genomic discoveries do not offer support for the Destiny Myth. Instead, as the example of the mixed-race twin sisters Millie and Marcia Biggs highlights, DNA affects people because of—not despite—the many social and environmental features of the world. Phrases like "genetic effects" or the "effects of DNA" can obscure this fact; people need new, clearer terminology for describing the ways that a person's DNA affects their life. Such changes will take time and require many voices to come together.[1] For now, however, it is crucial to keep in mind that these phrases do not refer to strictly genetic processes, per se, but instead often entail an intricate, context-specific relationship involving a person's DNA along with particular social and physical environments.

Chapter 3 unraveled the Race Myth, or the false belief that DNA divides humans into discrete and biologically distinct racial groups. More than one in five Americans has taken a genetic ancestry test to obtain estimates of the proportion of their DNA that falls into various geographic categories. Many people inaccurately conflate ancestry with race and therefore believe that these genetic tests challenge the idea that race is a sociopolitical construct that operates to benefit some and harm others. However, understanding what ancestry, at its core, is—the expansive Family Tree of all of humanity—helps distinguish it from race and racialization, and will help you interpret your genetic ancestry test results in an informed way. Processes of social construction often involve humans assigning meaning to the physical, natural, and biological features of the world; genetic ancestry tests take the expansive Family Tree of humanity and imbue it with social meaning.

Part 2 disentangled Daphne's and Sam's distinct perspectives on controversies related to genetic myths and genomic research. Chapter 4 explored the historical effects of genetic myths and their persistent place in our social world. The opening and closing dialogues in this chapter unpacked distinct ideas about the role of genetic myths in shaping an ever-evolving American society. Chapter 5 outlined debates in social genomics about how research gets used, including who decides its risks and benefits. This chapter also discussed the relationship between genetic myths and emerging genomic research, emphasizing the ways that such myths can shape the uses and interpretations of modern genomic tools. Chapter 6 described connections between DNA and social inequality, arguing that distinct understandings of social inequality sit at the core of debates and controversies on the subject. This chapter offers two distinct conceptualizations of social inequality and highlights how the differences between these two frameworks can lead people who share an interest in building a more equal world to talk past each other.

Finally, part 3 delved into a policy framework to respond to the growing availability and accuracy of polygenic scores. Procedures that were once mere objects of science fiction, like gene editing and polygenic embryo selection, are now technological realities. The ability to use technologies like polygenic scores as tools for social distinction—or worse, social hierarchy—requires reflection on the proper uses of these resources. Private industry is running away with applications of polygenic scores, and controversy surrounding DNA must not stifle the development and implementation of key policy guardrails. Chapter 7 considered the regulation of polygenic embryo selection, and chapter 8 outlined the regulation of direct-to-consumer genetic testing and polygenic-informed screening programs.

The promises and perils of polygenic scores are often distributed across different domains. While genetic myths threaten to reify and exacerbate persistent structural inequalities (like racial and socioeconomic inequality), novel DNA-based tools offer the ability to identify previously hidden axes of individual difference. The unevenness of the risk-benefit profile presented by human genomics research can breed contention, especially when it comes to the study of social and behavioral traits. The application of polygenic scores should be regulated in ways that avoid reinforcing and deepening structural inequalities.

————

In Buffalo, New York, the symbol for Sankofa adorns many of the concrete railings that line Cherry Street—the same railings that Kat Massey vigorously and successfully lobbied her state to erect.

Sankofa is an idea from the Akan Tribe of Ghana that literally means "to go back and retrieve"; it conveys the message that building a better future requires reflecting on the past. Kat's life's work, as well as her murder, should serve as a reminder: to successfully navigate the coming years of rapid scientific advances and the ensuing proliferation of genomic tools, society must hold firmly in its collective mind the hard lessons learned over the past few centuries regarding genetic myths.

———

SAM: *I think we both seriously underestimated how tough writing this book would be.*

DAPHNE: *Oh, absolutely. I'm not sure I would've been up for it if I'd known what it'd really entail—no offense.*

SAM: *None taken! I know exactly what you mean. In retrospect, a huge part of what has been so challenging is our disciplinary differences. We gather data and construct knowledge in different ways. We ask different types of research questions that often can't be answered using the same sort of evidence. When we started, I didn't speak your language, and you didn't speak mine.*

DAPHNE: *For sure. On the surface, it may seem like questions about the effects of genetic myths can be answered in the same way as questions about the effects of a person's DNA. In reality, the task of figuring out which DNA variants cause a person to live longer is just so different from identifying the factors that, for instance, led the eugenics movement to gain traction during the first half of the twentieth century.*

SAM: *Yeah, and I think a key tension is how "zoomed in" or "zoomed out" a question is. Questions about the effects of DNA tend to be zoomed in—here, the classic scientific toolkit works quite nicely. The world has billions of people, and researchers can think of each one of them as part of a broader experiment; then, they can make an analysis plan, gather a large enough* sample, and test their hypotheses. Questions about the effects of genetic myths are far more zoomed out—the unit of analysis is not the individual, but the society. In a sense, there's only a single observation for this latter kind of question.

DAPHNE: Sure—but we don't always need to gather a sample or run an experiment to create knowledge. Just by living their lives, people learn new things every day without conducting any kind of formal scientific study. History, not a randomized experiment, provides numerous examples of how genetic myths were implicated in a variety of social harms. Answers to expansive questions—like how genetic myths shape the way we perceive ourselves and others, or what the world today would look like without social genomics research—cannot be definitively answered using the tools of modern science. Does that mean we abandon the task of grappling with these questions altogether? No. We cannot let challenges to obtaining certain, definitive, scientific answers stop us from asking big, important questions about our history and our future.

SAM: In the end, we agree that both types of questions are important—both ways of making meaning are necessary and valid in their own way. I think it's all too easy for a bioethicist to dismiss a geneticist's research as too narrow and simplistic or for a geneticist to think that a bioethicist's work is ascientific or abstract.

DAPHNE: Definitely. Our communities tend to dismiss each other, and it worries me! Genomic technologies are rapidly advancing and becoming more accessible and affordable. Consumers need to know what exactly they are spending their money on so they can make informed decisions. Policymakers need to know what polygenic scores do, and do not, capture so they don't use them inappropriately. Schoolchildren need to learn about the complexities of human biology so they don't fall for common genetic myths. How can these changes occur if folks stay in their echo chambers?

SAM: *Finding ways to build—rather than burn—bridges is more important than ever. And it's only becoming more challenging to anticipate what the future may bring, and when. I mean, the impact of polygenic embryo selection is currently limited by our ability to harvest embryos. Today, prospective parents can usually select from a dozen or fewer. But before too long, scientists may be able to artificially produce many more embryos, which would really change the math on what's possible in terms of the expected change. And what about polygenic gene editing? It's conceivable that, one day, you could use CRISPR to increase the prevalence of DNA variants that are currently quite rare but nonetheless have large effects.*

DAPHNE: *So much is uncertain about the future and how genomic technologies will evolve. What I think is certain is that talking about and trying to anticipate the future is valuable. As we learned to communicate with each other better, putting different perspectives into conversation with one another became a key goal of our book. Ultimately, and despite a lot of frustrating moments, writing this book with you has been worth it. I can only hope that, in our deeply polarized world, more people undertake the hard work of trying to engage with those they disagree with. It sure wasn't easy, though. We had a really steep learning curve. What do you think you've learned from this project?*

SAM: *It was curiosity—and a bit of skepticism—regarding the technical aspects of polygenic scores that first got me interested in social genomics research . . . what exactly is "in" a polygenic score, how well do they actually work, and so on. So, in a sense, I've always been interested in questions regarding how polygenic scores can or cannot be used. Working with you and writing this book helped me to think about these issues from an ethical, rather than scientific, perspective. Now, I find myself spending more and more of my time wading through conceptual questions about DNA and polygenic scores, as well as translating genomics research to wider audiences. I think that's*

*a good thing—because our databases are growing, and new
studies are coming out every week. We need some folks with
scientific expertise to slow down, digest, and distill all the new
genomic discoveries and technologies. What about you—what
have you learned?*

DAPHNE: *I think I better understand how social genomics
research can help create tools that are useful to scientists and,
perhaps eventually, to society. Developing and improving a
dyslexia polygenic score, for instance, might lead to changes in
how children are diagnosed and treated for the condition.
To be clear, my views on the risks of social genomics haven't
changed much—but I also don't see all the researchers who use
DNA to study social and behavioral outcomes as villains or
people working directly in opposition to me. I think most of
the social genomics researchers I've interacted with have the
same goals as me: living in a world with less social inequality,
treating each human with respect and dignity, and trying to
prevent inappropriate, or premature, applications of polygenic
scores. We just have different visions of how to go about it.*

SAM: *Maybe part of why you initially saw social genomics
researchers as somewhat nefarious was the fact that talking
about DNA can be taboo—especially when it comes to social
and behavioral traits. Personally, I think when people view
DNA as forbidden or controversial, it can actually draw
unneeded attention to genomics research. There becomes an
incentive for media outlets to publish cute, but wrongheaded,
stories about DNA—like news headlines claiming we've cracked
the code and finally found the depression genes. If only it was
that simple! Of course, DNA data has its uses—but those uses
are, like most scientific advances, much more measured and,
frankly, sometimes a bit boring.*

DAPHNE: *Yeah, if I've learned anything from writing this book, it's
that everything is much more nuanced than we initially think it
is. For example, I now realize just how complicated it is to try to
cleanly distinguish social genomics from, say, medical genomics.*

It's no longer clear to me where social genomics starts and stops as an area of research.

SAM: *Agreed. I think there's this idea that knowing that DNA affects medical traits is totally uncontroversial and unproblematic. But when we're talking about nonmedical traits—and especially those that are socially valued or constructed—thinking about the role of DNA is very fraught. Is it really that simple though? What side of the line does a trait like ADHD fall on? A person experiencing the condition might be treated by a psychiatrist, but its ultimate consequences typically manifest in schools and workplaces. And, as society medicalizes ADHD more and more over time, where exactly it falls on the "medical-social spectrum" is changing.*

DAPHNE: *I think that's a key reason why we ended up writing a book on polygenic scores instead of just social genomics. We realized that there isn't a single way to draw a line between the medical and the social. In fact, the collective negotiation currently happening around how we should draw this line is an important part of the story in and of itself.*

SAM: *Totally, we're in the midst of a new societal conversation regarding what questions to prioritize for scientific exploration— and which applications are ethically appropriate at all.*

DAPHNE: *Given the many difficult conversations we'd already had, I was expecting more of the same when we started writing the policy chapters at the end. I mean, I think I nearly called it quits in the middle of writing chapter 6! But, once we reached the final few chapters of the book, we'd formed a foundation of shared understanding. We had learned to see things from each other's perspective—so it actually felt easier than I expected!*

SAM: *Yeah, we quickly agreed that regulation is lacking but necessary.*

DAPHNE: *All those earlier debates and disagreements helped us find something that we could both get behind: a framework for how polygenic scores might be used to help people without exacerbating structural inequalities.*

SAM: *Even as the world seemingly becomes more and more polarized, this experience with you has made me a bit optimistic: if we make the time and space, and just keep coming back, disputes that sometimes seem intractable actually begin to give way. This is so crucial because, at least when it comes to genomics, the questions and disputes are only going to become harder.*

DAPHNE: *Geez, we've really got our work cut out for us, don't we?*

Technical Appendix

The estimates of the expected gain from polygenic embryo selection presented in figures 6, 7, and 8 are derived using equation 1 provided in Karavani et al.'s work.[1] For East Asian, Indigenous American, and Sub-Saharan African ancestries, we follow Turley et al. and scale the expected gains for European ancestries down by the prediction factor identified in Martin et al.[2] See table A.1 for the specific genomic study (i.e., GWAS) and ancestral population multiplier used for each trait-year-ancestry estimate. Unless otherwise specified, the estimates presented in this book assume ten embryos. HR-inches are constructed using a standard deviation of 2.94 inches, which is the average of the estimated standard deviations of the male (3.05 inches) and female (2.83 inches) height distributions using Wave IV of the National Study of Adolescent to Adult Health. In the long run, if novel reproductive technologies allow for the generation of many more embryos, then the expected gains for a given polygenic score will increase. Similarly, if new genomic methods allow polygenic scores to incorporate information from rare variants, then it's possible that the theoretical maximum correlation would increase and the expected gain would be even larger.

Table A.1. Inputs into Calculations of Expected Change from Polygenic Embryo Selection

Trait	Genomic Study	Sample Size	Ancestry	Number of Embryos	Within-Family R²	Population Multiplier	Expected Change
Height	*Maximum* (h^2_{SNP} from Yengo et al. 2022)[a]	∞	European	10	0.51	1	2.46 inches
Height	Lango Allen et al. 2010,[b] Authors' calculations	183,727	European	10	0.029*	1	0.59 inches
Height	Yengo et al. 2022[a]	5,380,080	European	10	0.33	1	1.98 inches
Cardiovascular Disease	*Maximum* (h^2_{SNP} from Tcheandjieu et al. 2022)[c]	∞	European	10	0.34	1	2.01 HD-inches
Cardiovascular Disease	Khera et al. 2018,[d] Turley at al. 2021[e]	606,202	European	10	0.014	1	0.41 HD-inches
Cardiovascular Disease	Khera et al. 2018,[d] Turley et al. 2021[e]	606,202	European	4	0.014	1	0.32 HD-inches
Cardiovascular Disease	Schunkert et al. 2011[f]	86,995	European	10	0.001	1	0.11 HD-inches
Height	Yengo et al. 2022[a]	5,380,080	Indigenous American	10	0.33	0.63	1.28 inches
Height	Yengo et al. 2022[a]	5,380,080	East Asian	10	0.33	0.5	0.99 inches
Height	Yengo et al. 2022[a]	5,380,080	Sub-Saharan African	10	0.33	0.22	0.44 inches
Cardiovascular Disease	Khera et al. 2018,[d] Turley et al. 2021[e]	606,202	Indigenous American	10	0.014	0.63	0.26 HD-inches
Cardiovascular Disease	Khera et al. 2018,[d] Turley et al. 2021[e]	606,202	East Asian	10	0.014	0.5	0.21 HD-inches
Cardiovascular Disease	Khera et al. 2018,[d] Turley et al. 2021[e]	606,202	Sub-Saharan African	10	0.014	0.22	0.09 HD-inches

*This estimate is based on the results presented in Lango Allen et al. (see note b below). The study conducts GWAS on a sample of 183,727 individuals, and the resulting polygenic score yields a between-family correlation with observed height of $r = 0.36$. However, in order to determine the expected change from polygenic embryo selection, we need to know the correlation between polygenic score and height within families. Unfortunately, the publicly available summary statistics from Lango Allen and colleagues include p-values but not β-values, making it impossible to generate a polygenic score based on the results. However, a follow-up height GWAS which used a slightly expanded sample size of 253,288 individuals, from Wood et al. (see note g below), reports both p-values and β-values in its summary statistics. The results from this study produce a polygenic score that has a between-family correlation with observed height of $r = 0.54$. Using the results from Wood et al., we generate a polygenic score in a sample of 468 genotyped sibling pairs of European ancestries in the National Study of Adolescent to Adult Health. Phenotypic height is clinically measured at Wave IV, when study members were ages 24–32. We first residualize height on age and sex, and then estimate the correlation between polygenic score and height while controlling for family fixed effects. This procedure provides an estimated within-family correlation of $r = 0.26$. We use the relationship between the within- and between-family r values for the Wood et al. result to transform the Lango Allen et al.'s between-family $r = 0.36$ to an estimated within-family $r = 0.17$.

[a] Loïc Yengo, Sailaja Vedantam, Eirini Marouli, Julia Sidorenko, Eric Bartell, Saori Sakaue, Marielisa Graff et al. "A Saturated Map of Common Genetic Variants Associated with Human Height." *Nature* 610, no. 7933 (2022): 704–712.

[b] Hana Lango Allen, Karol Estrada, Guillaume Lettre, Sonja I. Berndt, Michael N. Weedon, Fernando Rivadeneira, Cristen J. Willer et al. "Hundreds of Variants Clustered in Genomic Loci and Biological Pathways Affect Human Height." *Nature* 467, no. 7317 (2010): 832–838.

[c] Catherine Tcheandjieu, Xiang Zhu, Austin T. Hilliard, Shoa L. Clarke, Valerio Napolioni, Shining Ma, Kyung Min Lee et al. "Large-scale Genome-wide Association Study of Coronary Artery Disease in Genetically Diverse Populations." *Nature Medicine* 28, no. 8 (2022): 1679–1692.

[d] Amit V. Khera, Mark Chaffin, Krishna G. Aragam, Mary E. Haas, Carolina Roselli, Seung Hoan Choi, Pradeep Natarajan et al. "Genome-wide Polygenic Scores for Common Diseases Identify Individuals with Risk Equivalent to Monogenic Mutations." *Nature Genetics* 50, no. 9 (2018): 1219–1224.

[e] Patrick Turley, Michelle N. Meyer, Nancy Wang, David Cesarini, Evelynn Hammonds, Alicia R. Martin, Benjamin M. Neale et al. "Problems with Using Polygenic Scores to Select Embryos." *New England Journal of Medicine* 385, no. 1 (2021): 78–86.

[f] Heribert Schunkert, Inke R. König, Sekar Kathiresan, Muredach P. Reilly, Themistocles L. Assimes, Hilma Holm, Michael Preuss et al. "Large-scale Association Analysis Identifies 13 New Susceptibility Loci for Coronary Artery Disease." *Nature Genetics* 43, no. 4 (2011): 333–338.

[g] Andrew R. Wood et al. "Defining the Role of Common Variation in the Genomic and Biological Architecture of Adult Human Height." *Nature Genetics* 46 (2014): 1173–1186. https://doi.org/10.1038/ng.3097.

NOTES

Preface

1. The G&SS group was organized by Ben Domingue and attendees included Amal Harrati, David Rehkopf, Genevieve Wojcik, Jeremy Freese, Larami Duncan, Marcus Feldman, Robert Willis, Shripad Tuljapurkar, and others. Three of the five members of Sam's dissertation committee were regular attendees of G&SS.

2. In a traditional adversarial collaboration, parties are refereed by a neutral arbiter who sets and manages expectations and monitors discussions to avoid later miscommunications. Each party designs an empirical experiment that they believe will prove their hypothesis to be the correct one. Then they come together to devise a joint experiment that ideally incorporates elements of each party's approaches; through this collaborative exercise, "adversaries" learn not only about how their counterpart thinks, but how they themselves think. Our joint experiment is, of course, this book—though it is worth noting that while adversarial collaboration was originally conceived for resolving empirical disputes, our disputes are not all empirical; we have conceptual and normative disagreements as well. Thus, this book is in many ways an adaptation of adversarial collaboration. Our neutral arbiter was Zoom; we recorded our conversations when things got especially heated or tricky to navigate. We consider yoga to be another surprising mediator; though we never took a class together, this shared hobby allowed each of us to escape and decompress when we wanted to bang our heads against a wall. For examples of our previous collaborations, see Daphne Martschenko, Sam Trejo, and B. W. Domingue. "Genetics and Education: Recent Developments in the Context of an Ugly History and an Uncertain Future." *AERA Open* 5, no. 1 (2019). https://journals.sagepub.com/doi/10.1177/2332858418810516, and Daphne Martschenko and Sam Trejo. "Ethical, Anticipatory Genomics Research on Human Behavior Means Celebrating Disagreement." *Human Genetics and Genomic Advances* 3, no. 1 (2022). https://doi.org/10.1016/j.xhgg.2021.100080.

3. "Make Science More Collegial: Why the Time for 'Adversarial Collaboration' Has Come." *Nature* 641, no. 8062 (May 6, 2025): 281–282. https://doi.org/10.1038/d41586-025-01379-3.

4. Daniel Kahneman. "Experiences of Collaborative Research." *American Psychologist* 58, no. 9 (2003): 723–730.

5. Catherine Bliss. *Social by Nature: The Promise and Peril of Sociogenomics*. Stanford, CA: Stanford University Press, 2018; Dalton Conley and Jason Fletcher. *The Genome Factor*. Princeton, NJ: Princeton University Press, 2017; Katherine P. Harden. *The Genetic Lottery: Why DNA Matters for Social Equality*. Princeton, NJ: Princeton University Press, 2021; Dalton Conley. *The Social Genome: The New Science of Nature and Nurture*. New York: Liveright, 2025.

6. For instance, at the outset of this book project, Sam already considered himself more moderate than many of the most outspoken members of the social genomics community. He hails from sociology which, unlike other social scientific disciplines such as economics and psychology, has long been interested in the role that genetic myths play in producing (and reproducing) social inequality. He doesn't think social genomics will revolutionize the social sciences or society—the promise of most technological advances are often overhyped, at least at first—but he sees the research as useful and believes the benefits outweigh the risks.

The phrase "murky middle" comes from Philip E. Tetlock and Gregory Mitchell. "Implicit Bias and Accountability Systems: What Must Organizations Do to Prevent Discrimination?" *Research in Organizational Behavior* 29 (2009): 3–38. https://doi.org/10.1016/j.riob.2009.10.002.

1. Two Acids

1. Robert Rosenthal and Lenore Jacobson. *Pygmalion in the Classroom: Teacher Expectation and Pupils' Intellectual Development*. New York: Holt, Rinehart and Winston, 1968.

2. Nicholas W. Papageorge, Seth Gershenson, and Kyung Min Kang. "Teacher Expectations Matter." *Review of Economics and Statistics* 102, no. 2 (2020): 234–251. https://doi.org/10.1162/rest_a_00838.

3. For an example of a past book about the effects of DNA and scientific developments at the intersection of genomics and the social sciences, see Dalton Conley and Jason Fletcher. *The Genome Factor*. Princeton, NJ: Princeton University Press, 2017. For an example of a past book about genetic myths and the ways in which they influence how researchers think about DNA and individual difference, see Catherine Bliss. *Social by Nature: The Promise and Peril of Sociogenomics*. Stanford, CA: Stanford University Press, 2018.

4. "How Diplomacy Helped to End the Race to Sequence the Human Genome." *Nature*, June 24, 2020.https://www.nature.com/articles/d41586-020-01849-w.

5. "Text of the White House Statements on the Human Genome Project." *New York Times*, June 27, 2000. https://archive.nytimes.com/www.nytimes.com/library/national/science/062700sci-genome-text.html.

6. Boyle, Evan A., Yang I. Li, and Jonathan K. Pritchard. "An Expanded View of Complex Traits: From Polygenic to Omnigenic." *Cell* 169, no. 7 (2017): 1177–1186.

7. In a 2021 paper published in *Nature Human Behaviour*, the term *polygenic index* was proposed as a possible alternative to *polygenic score* (see Joel Becker et al. "Resources Profile and User Guide of the Polygenic Index Repository." *Nature Human Behaviour* 5, no. 12 (2021): 1744–1758. doi: 10.1038/s41562-021-01119-3). This proposal was motivated by the idea that *polygenic index* may be "less likely to give the impression of a value judgement where one is not intended" than *polygenic score*. While, to our knowledge, the extent to which each term communicates normative sentiment in the context of human genomics has yet to be empirically explored, we are nevertheless sympathetic to the authors' thoughtfulness and concern. However, we ultimately decided to use polygenic score throughout the book for a couple of reasons. First, *What We Inherit* is intended for wider audiences; we believe that the term "score" is more accessible than "index" (which is used less frequently outside of academic settings). Second, while the use of polygenic index is gaining traction in research at the intersection of genomics and the social sciences, our book considers the use of new genomic tools for *both* medical and social traits; in the medical literature, it is our sense that polygenic score is still the most commonly used term.

8. Kathleen Mullan Harris et al. "Cohort Profile: The National Longitudinal Study of Adolescent to Adult Health (Add Health)." *International Journal of Epidemiology* 48, no. 5 (2019): 1415–1415k. doi: 10.1093/ije/dyz115.

9. Cornelius Rietvald et al. "GWAS of 126,559 Individuals Identifies Genetic Variants Associated with Educational Attainment." *Science* 340, no. 6139 (2013): 1467–1471. doi: 10.1126/science.1235488.

10. There has, of course, been ample debate about what exactly IQ tests capture and how they can (or should) be used. Stephen Jay Gould. *Mismeasure of Man*. New York: Norton, 1996; Ken Richardson. "What IQ Tests Test." *Theory & Psychology* 12, no. 3 (2002): 283–314; Stuart Ritchie. *Intelligence: All That Matters*. London: Hachette UK, 2015.

11. Aysu Okbay et al. "Polygenic Prediction of Educational Attainment within and between Families from Genome-Wide Association Analyses in 3 Million Individuals." *Nature Genetics* 54 (2022): 437–449. https://doi.org/10.1038/s41588-022-01016-z.

12. Genomic principal components are standard controls for population structure in the literature.

13. Rietvald et al., "GWAS of 126,559 Individuals Identifies Genetic Variants," 1467–1471.

14. Okbay et al., "Polygenic Prediction of Educational Attainment within and between Families," 437–449.

15. Despite this distinction, no hard-formed line exists to divide physical traits (e.g., height) from medical traits (e.g., diabetes) from social and behavioral traits (e.g., education). You need only to open up a dating app like Tinder to glimpse the intricate way meaning and value become layered upon even a seemingly innocuous trait like how tall one is. Moreover, many of the practical and conceptual limitations of polygenic scores apply to a wide range of traits that supersede the arbitrary boundaries among physical, medical, and behavioral outcomes.

16. Additionally, this book exclusively focuses on genomics (rather than, for example, epigenomics) because genomics research is more likely to become a vessel for genetic myths. While a person's genome is set at birth, the field of epigenomics studies the biological modifications that become layered on top of their DNA sequence in response to environmental stimuli. We focus on genomics, in particular, because the unchanging nature of DNA features prominently in genetic myths, and DNA-related evidence often lays the groundwork for arguments about the immutability of life outcomes (i.e., DNA is permanent, so the effects of DNA are permanent). Another reason we focus on genomics is that epigenomics is earlier in its development as a field, in part because epigenomic data is more expensive to collect and is cell-type specific.

17. Siddhartha Mukherjee. *The Gene: An Intimate History*. New York: Scribner, 2016.

18. We defined these myths in such a way to make them logically false. For this reason, presenting evidence to empirically undermine these myths is explicitly not a goal of this book. Further, we use the phrases "Destiny Myth" and "Race Myth," but there is extensive scholarship in fields such as sociology, history, philosophy, and science & technology studies that articulates these myths in other ways. The Race Myth is an example of *racial essentialism*, whereas the Destiny Myth is an illustration of *genetic determinism*. For more on genetic determinism and genetic racial essentialism, see: Troy Duster. *Backdoor to Eugenics*. New York: Psychology Press, 2003; Dorothy Nelkin and M. Susan Lindee. *The DNA Mystique*. Ann Arbor: University of Michigan Press, 2004; and Richard Dawkins. *The Extended Phenotype: The Long Reach of the Gene*. Oxford University Press, 2016. For a discussion of the politics that shape American beliefs and attitudes about genetic inheritance and genomic technologies today, see Jennifer Hochschild. *Genomic Politics: How the Revolution in Genomic Science Is Shaping American Society*. Oxford: Oxford University Press, 2021. Finally, it is important to note that the Destiny Myth and Race Myth are not practiced uniformly. Celeste Condit, a scholar in communication studies, illustrates how ideas about DNA difference or genetic "causes" are used in ways that serve people's preexisting beliefs and agendas. For more, see Celeste M. Condit. "Laypeople Are Strategic Essentialists, Not Genetic Essentialists." *Looking for the Psychosocial Impact of Genomic Information* 49, no. 1 (2019): S27–S37. https://doi.org/10.1002/hast.1014.

19. When we discuss "class" in this book, we are specifically referring to the social and economic circumstances into which someone was born (that is, their childhood socioeconomic status).

2. The Destiny Myth

1. "These Twins, One Black and One White, Will Make You Rethink Race." *National Geographic*, March 12, 2018. https://www.nationalgeographic.co.uk/people -and-culture/2018/04/these-twins-one-black-and-one-white-will-make-you -rethink-race.

2. Julia Macfarlane. "'One-in-a-million' Biracial Twins Won't Let Race Define Them: 'You Don't Always Have to Blend In.'" *Good Morning America*, March 12, 2018. https://www.goodmorningamerica.com/family/story/million-biracial-twins -define-blend-53681241.

3. "These Twins, One Black and One White, Will Make You Rethink Race."

4. We have deliberately conceptualized the Destiny Myth as unrelated to beliefs about the magnitude of a given trait's heritability. Instead, our formulation of the Destiny Myth is about the stability of the effects of DNA (and, so too, heritability) across place and time. We believe it is a myth to say that a trait with a high heritability in a given context means that there do not exist other—either actual or possible—contexts where that same trait has a low heritability.

5. The word "gene" first appeared in the Danish botanist Wilhelm Johannsen's 1905 book *The Elements of Heredity* to describe units of biological inheritance; its use predates the discovery of the DNA helix structure. The discovery of the DNA helix structure is most often credited to Rosalind Franklin, James Watson, and Francis Crick in the 1950s. The Swiss chemist Johann Friedrich Miescher discovered the chemical composition of DNA nearly a century earlier in 1869—the very year Galton published *Hereditary Genius*. For more information, see Ehud Lamm, Oren Harman, and Sophie J. Veigl. "Before Watson and Crick in 1953 Came Friedrich Miescher in 1869." *Genetics* 215, no. 2 (2020): 291–296. doi:10.1534/genetics.120.303195.

6. Francis Galton. *Hereditary Genius: An Inquiry into Its Laws and Consequences*. London: Macmillan, 1869.

7. Peter H. Lindert. "Unequal English Wealth since 1670." *Journal of Political Economy* 94 (1986): 1127–1162.

8. Full quote: "I acknowledge freely the great power of education and social influences in developing the active powers of the mind, just as I acknowledge the effect of use in developing the muscles of a blacksmith's arm, and no further. Let the blacksmith labour as he will, he will find there are certain feats beyond his power that are well within the strength of a man of herculean make, even although the latter may have led a sedentary life . . . There is a definite limit to the muscular powers of every man, which he cannot by any education or exertion overpass. This is precisely analogous to the experience that every student has had of the working of his mental powers." Galton. *Hereditary Genius*.

9. Christopher Jencks. "Heredity, Environment, and Public Policy Reconsidered." *American Sociological Review* 45, no. 5 (1980): 723–736; Jeremy Freese. "Genetics

and the Social Science Explanation of Individual Outcomes." *American Journal of Sociology* 114, no. 1 (2008): S1–S35. doi: 10.1086/592208.

10. Dalton Conley. *The Social Genome: The New Science of Nature and Nurture.* New York City: Liveright, 2025.

11. Oprah Winfrey. "Donald Trump on the Role Genetics Play in Success." April 25, 1988. *The Oprah Winfrey Show,.* https://www.oprah.com/own-oprahshow /donald-trump-on-the-role-genetics-play-in-success-video.

12. Racehorse theory originally emerged in the context of horse breeding. The theory posited that careful and intentional breeding of thoroughbreds would produce better horses. Eventually, early proponents of the eugenics movement applied the theory to their arguments about racial degeneration. Preeminent scientists and eugenicists such as Harry H. Laughlin, superintendent of research at the Eugenics Record Office at Cold Spring Harbor on Long Island, pored over horse pedigree charts in an attempt to understand biological inheritance and how the concept could be applied to bettering the human race. For more information, see Brian Tyrrell. "Bred for the Race: Thoroughbred Breeding and Racial Science in the United States, 1900–1940." *Historical Studies in the Natural Sciences* 45, no. 4 (2015): 549–576. https://doi.org/10.1525/hsns.2015.45.4.549. Also see Michael Kirk et al., producers. *The Choice 2016.* PBS Frontline, 2016. https://www.pbs.org/wgbh/frontline /documentary/the-choice-2016/.

13. Brandon Tensley. "The Dark Subtext of Trump's 'Good Genes' Compliment." CNN, September 22, 2020. https://www.cnn.com/2020/09/22/politics/donald -trump-genes-historical-context-eugenics/index.html; Michael Gold. "Trump's Long Fascination With Genes and Bloodlines Gets New Scrutiny." *New York Times,* December 22, 2023. https://www.nytimes.com/2023/12/22/us/politics/trump -blood-comments.html.

14. Nicole Narea. "Donald Trump's Long History of Enabling White Supremacy, Explained." *Vox,* November 29, 2022. https://www.vox.com/policy-and-politics /23484314/trump-fuentes-ye-dinner-white-nationalism-supremacy. Daphne also wrote an op-ed examining comments Trump made about "bad genes" during his 2024 re-election campaign: Daphne O. Martschenko. "The Alarming History Behind Trump's 'Bad Genes' Comments." The Hastings Center, October 15, 2024. https://www .thehastingscenter.org/the-alarming-history-behind-trumps-bad-genes-comments/.

15. Social genomics researchers are typically interested in identifying causal (and not merely correlational) relationships between DNA and social, behavioral, and health outcomes. Simply because such researchers are interested in these causal questions does not mean that existing genomics discoveries provide satisfying or definitive answers to them. In particular, a person's genome contains granular information about their place within a population. Naïve associations between DNA and outcomes among unrelated individuals are often environmentally confounded (i.e., there exists a correlation but no causation) as a result of population structure (see

John Novembre, Toby Johnson, et al. "Genes Mirror Geography within Europe." *Nature* 456, no. 2 (2008): 98–101. https://doi.org/10.1038/nature07331) and dynastic genetic effects (see Sam Trejo and Benjamin W. Domingue. "Genetic Nature or Genetic Nurture? Introducing Social Genetic Parameters to Quantify Bias in Polygenic Score Analyses." *Biodemography and Social Biology* 64, nos. 3–4 (2018): 187–215. doi: 10.1080/19485565.2019.1681257 and Augustine Kong et al. "The Nature of Nurture: Effects of Parental Genotypes." *Science* 359, no. 6374 (2018): 424–428. doi: 10.1126/science.aan6877). It is therefore difficult to know how well existing genome-wide association studies and subsequent polygenic scores capture the causal effects of DNA (see Lawrence J. Howe et al. "Within-Sibship Genome-Wide Association Analyses Decrease Bias in Estimates of Direct Genetic Effects." *Nature Genetics* 54, no. 5 (2022): 581–592. doi: 10.1038/s41588-022-01062-7; Tammy Tan et al. "Family-GWAS Reveals Effects of Environment and Matin on Genetic Associations" (2024). *medRxiv*; Ramina Sotoudeh, Sam Trejo, Arbel Harpak, and Dalton Conley. "Does Standard Adjustment for Genomic Population Structure Capture Direct Genetic Effects?" (2024): *bioRxiv*.) In this chapter, we focus on complexities regarding what it means conceptually for a DNA variant to causally affect an outcome. However, we acknowledge that existing polygenic scores likely encapsulate a combination of the effects of DNA and environmental confounding (especially in the case of social and behavioral traits).

16. The counterfactual framework is often also referred to as the *potential outcomes framework*, or the *Neyman-Rubin causal model*. The field of study surrounding the theory and application of experiments and quasi-experiments is often referred to as *causal inference*.

17. When running a randomized experiment is not possible, researchers instead turn to *natural* experiments. Such an approach leverages random exposures to naturally occurring stimuli and interventions (e.g., the rollout of policy changes or unplanned events like disasters). One well-known example of a natural experiment used the 1918 Spanish flu, a global pandemic that struck without warning and killed tens of millions of people worldwide, to understand how traumas experienced in utero can extend across an individual's lifespan. Researchers compared mortality outcomes of babies born just before the Spanish flu struck (the control group) to babies in utero at the time and therefore exposed to the flu (the treatment group). Strikingly, babies prenatally affected by the 1918 flu completed fewer years of schooling, earned less money, and were more likely to experience a physical disability than otherwise similar children born immediately before the pandemic. For further discussion, see Douglas Almond. "Is the 1918 Influenza Pandemic Over? Long-Term Effects of In Utero Influenza Exposure in the Post-1940 U.S. Population." *Journal of Political Economy* 114, no. 4 (2006): 672–712. https://doi.org/10.1086/507154. Here, the randomness of a child being born just before or during the 1918 Spanish

flu provides leverage to ethically and reasonably identify causal effects, even in a non-experimental setting.

18. Parent-child comparisons, like sibling comparisons, also increase the robustness of genetic effect estimates. Although, even within-family genomic analyses are not without their complications. In particular, they may provide a weighted average of individual-level effects that does not equal the population-level average treatment effect (see Carl Veller, Molly Przeworski, and Graham Coop. "Causal Interpretations of Family GWAS in the Presence of Heterogeneous Effects." *Proceedings of the National Academy of Sciences* 121, no. 38 (2024): e2401379121) and also may suffer from genetic confounding (see Carl Veller and Graham Coop. "Interpreting Population and Family-Based Genome-Wide Association Studies in the Presence of Confounding." *PLOS Biology* 22, no. 4 (2024): e3002511. doi:10.1101/2023.02.26 .530052).

19. Head Start is a comprehensive education, health, and nutrition program for children under the age of five—basically a preschool, pediatrician, and cafeteria all rolled in to one, available free of charge for eligible families. President Lyndon Johnson grew up on a ranch outside of Austin, Texas, and was the namesake of Sam's high school.

20. Jennifer Hochschild and Maya Sen. "Genetic Determinism, Technology Optimism, and Race: Views of the American Public." *ANNALS of the American Academy of Political and Social Science* 661, no. 1 (2015): 160–180. https://doi.org/10.1177 /0002716215587875.

21. Charles Murray. *Human Diversity: The Biology of Gender, Race, and Class.* New York: Grand Central Publishing, 2020.

22. In reference to his desire to end child support payments to unwed mothers, Murray once cheerfully said, "I don't want society to say to her, 'You made a mistake' . . . I want society to say, 'You did wrong'" (Jason Deparle. "Daring Research or 'Social Science Pornography'?: Charles Murray." *New York Times*, October 9, 1994. https://www.nytimes.com/1994/10/09/magazine/daring-research-or -social-science-pornography-charles-murray.html).

23. Full quote: "That means polygenic scores will offer social scientists something they've never had before: a secure place to stand in assessing what is innate and what is added by the environment" (Charles Murray. "Genetics Will Revolutionize Social Science." *Wall Street Journal*, January 27, 2020. https://www.wsj.com/articles /genetics-will-revolutionize-social-science-11580169106).

24. Some may argue that, while there do exist environmentally mediated effects of DNA, there also exist "direct" effects of DNA that are not environmentally mediated. However, this idea is subtly yet deeply fraught. In statistics and the social sciences, a causal pathway is direct if it does not operate through a given finite set of mediating pathways. For more information, see Judea Pearl, and Dana Mackenzie. *The Book of Why: The New Science of Cause and Effect.* New York: Basic Books, 2018.

(This is different from the use of the term in genomics, where "direct genetic effect" describes the causal effect of an organism's own DNA on that same organism's traits.) Importantly, a genetic effect cannot be direct in and of itself, but only with respect to a specific set of mediating variables. Let us consider a single candidate mediator: diet. Some DNA variants may affect lifespan indirectly via first influencing a person's diet. Other DNA variants that affect lifespan, irrespective of diet, can be said to have a direct effect on lifespan. However, the "environment" is a catchall term used to refer to all aspects of a person's life besides their DNA. When we define "environment" as the mediator, we subtly insert all possible nongenetic variables as mediators. In such a case, no possible causal pathway remains for so-called direct effects to operate through. In this sense, all genetic effects are environmentally mediated. For more information, see Callie H. Burt. "Challenging the Utility of Polygenic Scores for Social Science: Environmental Confounding, Downward Causation, and Unknown Biology." *Behavioral Brain Science* 46 (2022): e207. doi: 10.1017/S0140525X22001145; and Sam Trejo and Daphne O. Martschenko. "Beware of the Phony Horserace Between Genes and Environments." *Behavioral Brain Science* 46 (2023): e228. doi: 10.1017/S0140525X22002485.

3. The Race Myth

1. The Lovings' marriage license described Mildred as simply "Indian" (omitting any mention of her Black heritage), but in her letter to the ACLU five years later Mildred described herself as "Negro" and "Indian." Then, toward the end of her life, Mildred returned to identifying herself as only "Indian." Sally Jacobs. "50 Years Later, The Couple at The Heart of Loving v. Virginia Still Stirs Controversy." GBH News, June 11, 2017. https://www.wgbh.org/news/2017/06/11/news/50-years-later-couple-heart-loving-v-virginia-still-stirs-controversy.

2. General Assembly. "'An Act for Suppressing Outlying Slaves.' (1691)." *Encyclopedia Virginia*. accessed June 2, 2024. https://encyclopediavirginia.org/entries/an-act-for-suppressing-outlying-slaves-1691/.

3. It wasn't until 1878 that the Virginia Assembly decided to punish both partners. For more information, see "Loving v. Virginia (1967)." *Encyclopedia Virginia*, accessed November 24, 2024. https://encyclopediavirginia.org/entries/loving-v-virginia-1967/.

4. "Loving v. Virginia." *Oyez*, accessed June 2, 2024. https://www.oyez.org/cases/1966/395.

5. Richard Loving was killed by a drunk driver on June 20, 1975. Mildred lost an eye in that crash but lived until 2008.

6. "Richard Perry LOVING et ux., Apellants, v. COMMONWEALTH OF VIRGINIA." Cornell Law School: Legal Information Institute, accessed June 2, 2024. https://www.law.cornell.edu/supremecourt/text/388/1.

7. Hugo L. Black, the youngest of eight children, grew up poor and was largely self-educated; in the early 1920s, he became a member of the Robert E. Lee Klan No. 1 in Birmingham, Alabama. He resigned from the KKK in 1925 and became a US senator (D-AL). He was appointed to the US Supreme Court in 1937 and stated during his confirmation hearings: "Before becoming a Senator I dropped the Klan. I have had nothing to do with it since that time. I abandoned it. I completely discontinued any association with the organization." Justice Black was a member of the Democratic Party and a passionate New Dealer; he was known to keep a copy of the US Constitution in his coat pocket.

8. Marisa Peñaloza. "'Illicit Cohabitation': Listen to 6 Stunning Moments From Loving V. Virginia." NPR, June 12, 2017. https://www.npr.org/2017/06/12/532123349/illicit-cohabitation-listen-to-6-stunning-moments-from-loving-v-virginia.

9. Alfaro, Mariana. "Republican Sen. Mike Braun Says Supreme Court Should Leave Decisions on Interracial Marriage, Abortion to the States." *Washington Post.* March 22, 2022.

10. Joe Hernandez. "23andMe Is Filing for Bankruptcy. Here's What It Means for Your Genetic Data." NPR, March 24, 2025, sec. Technology. https://www.npr.org/2025/03/24/nx-s1-5338622/23andme-bankruptcy-genetic-data-privacy.

11. Wendy D. Roth and Biorn Ivemark. "Genetic Options: The Impact of Genetic Ancestry Testing on Consumers' Racial and Ethnic Identities." *American Journal of Sociology* 124, no. 1 (2018): 150–184; Alondra Nelson. *The Social Life of DNA: Race, Reparations, and Reconciliation After the Genome.* Boston, MA: Beacon Press, 2016; Ariela Schachter, René D. Flores, and Neda Maghbouleh. "Ancestry, Color, or Culture? How Whites Racially Classify Others in the U.S." *American Journal of Sociology* 126, no. 5 (2021): 1220–1263. https://doi.org/10.1086/714215.

12. Aaron Panofsky and Joan Donovan. "Genetic Ancestry Testing among White Nationalists: From Identity Repair to Citizen Science." *Social Studies of Science* 49, no. 5 (2019): 653–681. https://doi.org/10.1177/0306312719861434.

13. Amy Harmon. "Seeking Ancestry in DNA Ties Uncovered by Tests." *New York Times*, April 12, 2006. https://www.nytimes.com/2006/04/12/us/seeking-ancestry-in-dna-ties-uncovered-by-tests.html.

14. Alice B. Popejoy et al. "Clinical Genetics Lacks Standard Definitions and Protocols for the Collection and Use of Diversity Measures." *American Journal of Human Genetics* 107, no. 1 (2020): 72–82. doi: 10.1016/j.ajhg.2020.05.005.

15. M. Roy Wilson, Sarah H. Beachy, and Samantha N. Schumm, eds. *Rethinking Race and Ethnicity in Biomedical Research.* Washington, DC: National Academies Press, 2024. https://doi.org/10.17226/27913; *Using Population Descriptors in Genetics and Genomics Research: A New Framework for an Evolving Field.* Washington, DC: National Academies Press, 2023. https://doi.org/10.17226/26902.

16. Off-screen, Danny has lived an intense and constantly evolving life, occupying real-life roles of child-abuse victim, convicted drug dealer, heroin addict, professional gardener, movie star, four-time divorcee, and, most recently, LA restaurant mogul.

17. Except in rare cases, humans have exactly two biological parents from whom they inherit DNA. However, mitochondrial donation procedures now enable a baby to inherit DNA from *three* different individuals. For more information, see Ian Sample. "First UK Baby with DNA from Three People Born After New IVF Procedure." *The Guardian*, May 9, 2023. https://www.theguardian.com/science/2023/may/09/first-uk-baby-with-dna-from-three-people-born-after-new-ivf-procedure.

18. Technically speaking, Daphne and Sam actually share the same *trunk* of the family tree—there are not roots, per se, as the genealogical tree extends back to chimpanzees and beyond. We use roots, however, because our focus is only on the human species. But, in truth, when you go back through the thousands of generations, even the analogy of the human species' genealogy resembling a family tree begins to break down—we're left with an expansive and high-dimensional interconnected web of life.

19. We use the term familial ancestry to refer to the academic concept of *genealogical* ancestry, which is related to (but distinct from) *genetic* ancestry—a term that refers to the stochastic biological inheritance of DNA through humanity's expansive genealogical network. See Iain Mathieson and Aylwyn Scally. "What Is Ancestry?" *PLOS Genetics* 16, no. 3 (2020): e1008624. https://doi.org/10.1371/journal.pgen.1008624; alternatively, an informative video on the concept of genetic ancestry can be found at the following link: https://www.youtube.com/watch?v=t9clljkF31Y.

20. Sam Trejo and Klint Kanopka. "Using the Phenotype Differences Model to Identify Genetic Effects in Samples of Partially Genotyped Sibling Pairs." *Proceedings of the National Academy of Sciences* 121, no. 49 (2024): e2405725121.

21. Mosby Woods Elementary School has since been renamed Mosaic Elementary School. The decision to change the school's name came with support from descendants of the school's namesake, Colonel John S. Mosby. The descendants sent a letter to the school board to request that Fairfax County rename the school in the interest of "maintaining an inclusive environment for all students." For more information, see Scott Fields. "Fairfax School Board Chooses New Name for Mosby Woods." *Tysons Reporter*, February 19, 2021. https://www.tysonsreporter.com/2021/02/19/fairfax-county-school-board-chooses-new-name-mosby-woods/.

22. Liliane Cambraia Windsor, Eloise Dunlap, and Andrew Golub. "Challenging Controlling Images, Oppression, Poverty and Other Structural Constraints: Survival Strategies Among African American Women in Distressed Households." *Journal of African American Studies (New Brunswick)* 15, no. 3 (2011): 290–306. doi: 10.1007/s12111-010-9151-0.

23. Michael Davern et al. "General Social Survey 1972–2024. [Machine-readable data file]." GSS Data Explorer, accessed June 2, 2024. gssdataexplorer.norc.org.

24. Nikki Khanna. "'If You're Half Black, You're Just Black': Reflected Appraisals and the Persistence of the One Drop Rule." *Sociological Quarterly* 51, no. 1 (2010): 96–121. https://doi.org/10.1111/j.1533-8525.2009.01162.x; Charles H. Cooley. *Human Nature and the Social Order.* New York: Routledge, 1902; Nancy Lopez et al. "What's Your 'Street Race'? Leveraging Multidimensional Measures of Race and Intersectionality for Examining Physical and Mental Health Status among Latinxs." *Sociology of Race and Ethnicity* 4, no. 1 (2017): 49–66. https://doi.org/10.1177 /2332649217708798.

25. Michael Omi and Howard Winant. *Racial Formation in the United States.* New York: Routledge, 2014; Andreas Wimmer. *Ethnic Boundary Making: Institutions, Power, Networks.* New York: Oxford University Press, 2013.

26. Aliya Saperstein, Jessica M. Kizer, and Andrew M. Penner. "Making the Most of Multiple Measures: Disentangling the Effects of Different Dimensions of Race in Survey Research." *American Behavioral Scientist* 60, no. 4 (2015): 519–537. https:// doi.org/10.1177/0002764215613399.

27. Aliya Saperstein and Andrew M. Penner. "Racial Fluidity and Inequality in the United States." *American Journal of Sociology* 118, no. 3 (2012): 676–727. https:// doi-org.ezproxy.princeton.edu/10.1086/667722; Wendy Roth. "The Multiple Dimensions of Race." *Ethnic and Racial Studies* 39, no. 8 (2016): 1310–1338. https://doi .org/10.1080/01419870.2016.1140793.

28. Karen Gringsby Bates. "'A Chosen Exile': Black People Passing in White America." NPR, October 7, 2014. https://www.npr.org/sections/codeswitch/2014 /10/07/354310370/a-chosen-exile-black-people-passing-in-white-america.

29. Rachel Dolezal, who has since changed her name to Nkechi Amare Diallo, became the center of a public scandal for identifying as Black when both her parents are racially classified and identify themselves as White (see Denene Millner. "Why Rachel Dolezal Can Never Be Black." NPR, March 3, 2017. https://www.npr.org /sections/codeswitch/2017/03/03/518184030/why-rachel-dolezal-can-never-be -black). Before controversy ensued, Dolezal was the president of the Spokane, Washington chapter of the National Association for the Advancement of Colored People (NAACP); she also attended Howard University, a historically Black university. Interestingly, Dolezal was raised in a family with Black adopted siblings.

30. Eric Foner. *Reconstruction: America's Unfinished Revolution, 1863–1877.* New York: Harper & Row, 1988.

31. Social scientists use the term "racial triangulation" to describe the complex social process that racializes Asian Americans as subordinate to and distinct from White Americans while simultaneously as superior to and distinct from Black Americans. See Claire Jean Kim. "The Racial Triangulation of Asian Americans." *Politics & Society* 27, no. 1 (1999): 105–138. https://doi.org/10.1177/0032329299027001005.

32. James W. Loewen. *The Mississippi Chinese: Between Black and White.* Long Grove, IL: Waveland Press, 1988. The 2025 film *Sinners* (starring Daphne's celebrity

crush, Michael B. Jordan) depicts the intertwined and complex relationship between Black and Chinese Americans in the Jim Crow South (see Erika Hayasaki. "The Real History of the Complex Relationship Between Chinese and Black Americans in the Mississippi Delta." *Smithsonian Magazine*, May 13, 2025. https://www.smithsonian mag.com/history/the-real-history-of-the-complex-relationship-between-chinese -and-black-americans-in-the-mississippi-delta-180986615/).

33. The etymology of the English word "race" is disputed. The *Oxford English Dictionary*'s earliest inclusion of "race" occurred in the sixteenth century as a noun signifying "a family" or "the posterity of a common ancestor." For more information, see Bronwen Douglas. "Climate to Crania: Science and the Racialization of Human Difference." In *Foreign Bodies: Oceania and the Science of Race 1750–1940*, edited by Bronwen Douglas and Chris Ballard, 33–96. Canberra: ANU Press, 2008.

34. As western empires sought to expand using unpaid, dehumanizing, and forced labor, those in power commonly argued that slaves were biologically suited to endure the labor and conditions of slavery. For instance, in the mid-nineteenth century, Africans who were brought over to the United States supposedly uniquely suffered from certain diseases that justified their enslavement as a medical necessity. One such illness was *drapetomania*—a mental condition that caused slaves to "irrationally" run away from their masters. Another was *dysaesthesia aethiopica*, a kind of lethargy that plagued Africans who were neither enslaved nor overseen by Whites. Samuel A. Cartwright, the Louisiana physician, who in 1851 first came up with these fictitious diseases, prescribed the cure of leather anointment and a leather strap. For more information, see Andrew Curran. "Facing America's History of Racism Requires Facing the Origins of 'Race' as a Concept." *Time*, July 10, 2020. https://time .com/5865530/history-race-concept/.

35. Alondra Nelson. "Weapons for When Bigotry Claims Science as Its Ally." *Nature*, September 7, 2020. https://www.nature.com/articles/d41586-020-02546-4.

36. Thomas Jefferson. *Notes on the State of Virginia*. Boston: Lilly and Wait, 1832.

37. Madison Grant. *The Passing of the Great Race: Or, The Racial Basis of European Ancestry*. New York: Scribner, 1921.

38. Ta-Nehisi Coates. *Between the World and Me*. New York: Spiegel & Grau, 2015.

39. Catherine Bliss. *Race Decoded: The Genomic Fight for Social Justice*. Stanford, CA: Stanford University Press, 2012.

40. Elle Hunt. "'Your father's not your father': When DNA Tests Reveal More Than You Bargained For." *The Guardian*, September 18, 2018. https://www .theguardian.com/lifeandstyle/2018/sep/18/your-fathers-not-your-father-when -dna-tests-reveal-more-than-you-bargained-for.

41. Modern humans split off from archaic humans approximately 300,000 years ago (though interbreeding occurred far more recently), and the typical length of a human generation is approximately 29 years. See Carina M. Schlebusch et al. "Southern African Ancient Genomes Estimate Modern Human Divergence to 350,000 to

260,000 Years Ago." *Science* 358, no. 6363 (2017): 652–655. doi: 10.1126/science. aao6266; Pontus Skoglund and Iain Mathieson. "Ancient Genomics of Modern Humans: The First Decade." *Annual Review of Genomics and Human Genetics* 19 (2018): 381–404. doi: 10.1146/annurev-genom-083117-021749. Additionally, Sam's favorite article from the satirical newspaper *The Onion* makes light of the role of location plays in mate selection (see "18-Year-Old Miraculously Finds Soulmate in Hometown." *The Onion*, February 20, 2002. https://www.theonion.com/18-year-old -miraculously-finds-soulmate-in-hometown-1819566332).

42. Waldo R. Tobler. "A Computer Movie Simulating Urban Growth in the Detroit Region." *Economic Geography* 46 (1970): 234–240. https://www.jstor.org /stable/143141.

43. Nonetheless, there was likely still more historical migration and population mixing than scientists believed prior to the genomic era. For instance, in the Bronze Age (3000 BCE), the remains of two fifth-degree relatives were found 1400 km apart. See David Reich. *Who We Are and How We Got Here: Ancient DNA and the New Science of the Human Past.* New York: Pantheon Books, 2018; Harald Ringbauer et al. "Accurate Detection of Identity-by-Descent Segments in Human Ancient DNA." *Nature Genetics* 56 (2024): 143–151. https://doi.org/10.1038/s41588-023 -01582-w.

Such migrations often produce *genetic admixture*, which occurs when two individuals from distinct branches of the Family Tree reproduce, resulting in a child with DNA sequences from both branches. Both Sam and Daphne are the result of admixture, a phenomenon that holds a kind of mystical beauty: two branches of the Family Tree, which shared roots but grew apart, reuniting after countless generations—as if childhood friends reconnecting in adulthood. Some, however, argue that many uses of the term admixture are misleading, as it incorrectly implies that there are, in fact, discrete ancestry groups.

44. John Novembre et al. "Genes Mirror Geography within Europe." *Nature* 456 (2008): 98–101. https://doi.org/10.1038/nature07331.

45. Castes in India are another example of a set of groups that historically lived geographically near one another but nevertheless occupy relatively distinct branches of the Family Tree. A recent genomic study found strong evidence of endogamy— that within-caste marriage and mating patterns have persisted in India for thousands of years. For more information, see David Reich, Kumarasamy Thangaraj, Nick Patterson, Alkes L. Price, and Lalji Singh. "Reconstructing Indian Population History." *Nature* 461, no. 7263 (2009): 489–494. doi: 10.1038/nature08365.

46. Shamam Waldman et al. "Genome-Wide Data from Medieval German Jews Show that the Ashkenazi Founder Event Pre-Dated the 14th Century." *Cell* 185, no. 25 (2022): 4703–4716.e16. doi: 10.1016/j.cell.2022.11.002. Unlike the other major Abrahamic religions, Judaism historically has largely been a non-missionary religion—either one was born Jewish by having a Jewish mother, or they were not;

this is one reason why there can be particular vitriol toward non-Jewish women, disparagingly referred to as *shiksas* in Yiddish, who threaten to steal away Jewish men from the faith.

47. Marissa Thompson, Sam Trejo, A. J. Alvero, and Daphne Martschenko. "'They Have Black in Their Blood': Exploring How Genetic Ancestry Tests Affect Racial Appraisals and Classifications." (2023). *SocArXiv*. https://doi.org/10.31235/osf.io/8tnrk.

48. Dorothy Roberts. *Fatal Invention: How Science, Politics, and Big Business Re-Create Race in the Twenty-First Century*. New York: New Press, 2011.

49. One calculation, for instance, estimated that the average number of fifth cousins that a typical British person had was 17,300 (see "Cousin Statistics." International Society of Genetic Genealogy. Accessed May 5, 2024. https://isogg.org/wiki/Cousin_statistics)! In addition, a key challenge of using genomic data to infer the geographic location of recent ancestors is that we usually do not know where various populations lived historically (though studies of ancient DNA may provide some evidence); for this reason, in practice, genetic ancestry tests usually utilize the *contemporary* geographic location (and genetic information) of various present-day populations.

50. A key figure in the Mexican-American War was Antonio López de Santa Anna. Santa Anna ruled Mexico as president, dictator, and military commander intermittently during a span of more than two decades. In a Napoleonic fashion, Santa Anna was at one point deposed and sent to an island exile only to return and seize power once again. Sam's great-grandmother was born Margarita Santa Anna, leading Sam's grandfather (Margarita's son) to become fascinated by the possibility that he was himself descended from the extravagant Mexican leader.

51. Donald Trump. "Restoring Truth and Sanity to American History." Executive Order 14253, March 31, 2025. https://www.whitehouse.gov/presidential-actions/2025/03/restoring-truth-and-sanity-to-american-history/.

52. See also Jiannbin Lee Shiao et al. "The Genomic Challenge to the Social Construction of Race." *Sociological Theory* 30, no. 2 (2012): 67–88. https://doi.org/10.1177/0735275112448053; Jeremy Freese, Ben Domingue, Sam Trejo, Kamil Sicinski, and Pamela Herd. "Problems with a Causal Interpretation of Polygenic Score Differences Between Jewish and Non-Jewish Respondents in the Wisconsin Longitudinal Study" (2019). *SocArXiv*; Charles Murray. *Human Diversity: The Biology of Gender, Race, and Class*. New York: Grand Central Publishing, 2020.

53. The full quote from *Human Diversity* is: "A geneticist can say to the orthodox, 'Give me a large random sample of SNPs in the human genome, and I will use a computer algorithm, blind to any other information about the subjects, that matches those subjects closely not just to their continental ancestral populations, but, if the random sample is large enough, to subpopulations within continents that correspond to ethnicities.' If race and ethnicity were nothing but social constructs, that

would be impossible. It's actually a sure bet." However, the social constructivist view of race is entirely compatible with the existence of a correlation between DNA and racial identity. For instance, the sociologist Ann Morning describes how processes of racial identification and classification "incorporate information (alongside beliefs) about individuals' phenotypic characteristics or geographic origins" and therefore can be "informed by (or correlated with) biology." For more information, see Ann Morning. "Does Genomics Challenge the Social Construction of Race?" *Sociological Theory* 32, no. 3 (2014): 189–207. https://doi.org/10.1177/0735275114550881.

4. The Effects of Genetic Myths

1. Ruha Benjamin. "Prophets and Profits of Racial Science." *Kalfou A Journal of Comparative and Relational Ethnic* 5, no. 1 (2018): 41–53. http://dx.doi.org/10.15367/kf.v5i1.198.

2. High School Bioethics Curriculum Project. "Chapter 2 Carrie Buck & The Lynchburg State Colony." Georgetown University Kennedy Institute of Ethics, accessed June 2, 2024. https://highschoolbioethics.georgetown.edu/units/cases/unit4_2.html.

3. Jennifer Schmidt et al. "Emma, Carrie, Vivian: How a Family Became a Test Case for Forced Sterilizations." NPR, April 23, 2018.

4. Harry H Laughlin. "The Legal Status of Eugenical Sterilization." *Buck v Bell Documents*, January 2009. http://readingroom.law.gsu.edu/buckvbell/79. Harry Hamilton Laughlin, who was superintendent of the Eugenics Record Office through the 1920s, classified Carrie a member of the "shiftless, ignorant, and worthless class of anti-social whites." For more, see Dorothy Roberts. *Killing the Black Body: Race, Reproduction, and the Meaning of Liberty*. New York: Pantheon Books, 1997.

5. A. F. Tredgold. "The Feeble-Minded—A Social Danger." *Eugenics Review* 1, no. 2 (1909): 97–104. https://www.ncbi.nlm.nih.gov/pmc/articles/PMC2986626/.

6. Stephen Jay Gould. "Carrie Buck's Daughter." Scholarship Repository: The University of Minnesota Law School, 1985. https://scholarship.law.umn.edu/concomm/1015.

7. "The Supreme Court Ruling that Led to 70,000 Forced Sterilizations." NPR, March 7, 2016. https://www.npr.org/sections/health-shots/2016/03/07/469478098/the-supreme-court-ruling-that-led-to-70-000-forced-sterilizations.

8. The closest thing to overturning *Buck v. Bell* occurred during *Skinner v. Oklahoma*. The case, which was decided in 1942, outlawed sterilization as a punitive measure and produced a legal quandary sufficient to discourage the use of state involuntary sterilization laws. Despite *Skinner v. Oklahoma*, today, scores of women—many of whom are immigrants and racial minorities—are still illegally coerced into receiving sterilization procedures. For more details, see Maya Manian. "Immigration Detention and Coerced Sterilization: History Tragically Repeats

Itself." American Civil Liberties Union, September 29, 2020. https://www.aclu.org /news/immigrants-rights/immigration-detention-and-coerced-sterilization-history -tragically-repeats-itself. Or see US Congress. Senate. Committee on Homeland Security and Governmental Affairs. *Medical Mistreatment of Women in ICE Detention.* November 15, 2022. S. Rep, 1–103. https://www.hsgac.senate.gov/wp-content /uploads/imo/media/doc/2022-11-15%20PSI%20Staff%20Report%20-%20Medical%20Mistreatment%20of%20Women%20in%20ICE%20Detention.pdf.

9. S. Selden. "Transforming Better Babies into Fitter Families: Archival Resources and the History of the American Eugenics Movement, 1908–1930." *Proceedings of the American Philosophical Society* 149 (2005): 199–225; Francine Uenuma. "'Better Babies' Contests Pushed for Much-Needed Infant Health but Also Played into the Eugenics Movement." *Smithsonian Magazine,* January 17, 2019. https://www .smithsonianmag.com/history/better-babies-contests-pushed-infant-health-also -played-eugenics-movement-180971288/.

10. Susanna Speier. "Beauty Pageants and the Misunderstanding of Evolution Meet. . . . Again." *Scientific American* blog, June 29, 2011. https://blogs.scientific american.com/guest-blog/beauty-pageants-and-the-misunderstanding-of-evolution -meet-again/.

11. Nathaniel Comfort. *The Science of Human Perfection: How Genes Became the Heart of American Medicine.* New Haven, CT: Yale University Press, 2012; Alexandra Minna Stern. *Eugenic Nation: Faults and Frontiers of Better Breeding in Modern America.* Berkeley and Los Angeles: University of California Press, 2015; Troy Duster. *Backdoor to Eugenics.* New York: Routledge, 2003.

12. Karl O. Knausgaard. "Anders Breivik's Inexplicable Crime." *The New Yorker,* May 18, 2015. https://www.newyorker.com/magazine/2015/05/25/the -inexplicable.

13. Victoria Klesty. "Norwegian Killer Breivik Begins Parole Hearing with Nazi Salute." Reuters, January 18, 2022. https://www.reuters.com/world/europe/mass -killer-breiviks-parole-hearing-begin-tuesday-norway-2022-01-17.

14. "Investigative Report on the Role of Online Platforms in the Tragic Mass Shooting in Buffalo on May 14, 2022." Office of the New York State Attorney General Letitia James, October 18, 2022. https://ag.ny.gov/sites/default/files/buffalo shooting-onlineplatformsreport.pdf.

15. Camus believed that, while it would be absurd to think that different races do not exist, whether there were biological differences between them remained an open question. Camus was resolute in his conviction that race was a social and cultural concept but was uncertain as to whether there were scientific explanations for racial differences. Tarrant, like Camus, was more subtle in his discussion of biological differences between racial groups. He wrote about European heritage being a combination of genes, culture, and language, but did not describe Muslim individuals as biologically or genetically distinct from White individuals.

16. We do not cite their screeds because doing so would add to the very attention they covet.

17. "The Supreme Court Ruling that Led to 70,000 Forced Sterilizations."

18. In a nationally representative *Education Week* survey of 1,333 teachers conducted in October 2019, 29% said that genetics are somewhat to extremely significant in explaining academic gaps between Black students and White students. For more information, see Christina A. Samuels. "Who's to Blame for the Black-White Achievement Gap?" *Education Week*, January 7, 2020. https://www.edweek.org /teaching-learning/whos-to-blame-for-the-black-white-achievement-gap/2020/01. A related study, which used a list experiment to reduce social desirability bias, found that one in five Americans attribute Black-White economic inequality to genetic differences between the races. For further reading, see Ann Morning, Hannah Brückner, and Alondra Nelson. "Socially Desirable Reporting and the Expression of Biological Concepts of Race." *Du Bois Review: Social Science Research on Race* 16, no. 2 (2019): 439–455. doi:10.1017/S1742058X19000195; Seth Gershenson, Stephen B. Holt, and Nicholas Papageorge. "Who Believes In Me? The Effect of Student–Teacher Demographic Match on Teacher Expectations." *Economics of Education Review* 52 (2016): 209–224.

19. Daphne O. Martschenko. "The New Borderland: A Mixed-Methods Examination of Teacher Perceptions of Intelligence, Race, and Socioeconomic Status in Relation to Behavior Genetics." (PhD thesis, University of Cambridge, 2019). doi:10.17863/CAM.40448.

5. The Effects of Genomic Research

1. Tom Clements. "I fell down the rabbit hole of alt-right propaganda and this is what I learned." *The Independent*, September 5, 2019. https://www.independent.co .uk/voices/alt-right-white-nationalist-richard-spencer-jonathan-bowden-nick -griffin-bnp-a9091376.html.

2. The Molyneux quote is from Episode 2768, "Collective Guilt for Fun and Profit," appearing at approximately 26 minutes ("Collective Guilt for Fun and Profit— Saturday Call-In Show August 9th, 2014." *Freedomain Radio*. August 11, 2014. Podcast, 2:27:58. https://fdrpodcasts.com/2768/collective-guilt-for-fun-and-profit-saturday -call-in-show-august-9th-2014). Near the start of a different video (Tullius Cicero. "Clips of Stefan Molyneux Promoting White Supremacism and Segregation." YouTube Video, 24:27. August 23, 2016. https://www.youtube.com/watch?v=XOi-5D9YcKOI), Molyneux more clearly articulates the same idea: "Looking at human beings as one species is not biologically valid—we are a variety of subspecies."

3. "Will Genius Be Genetically Engineered?" May 28, 2017. Podcast, 1:32:16. https://fdrpodcasts.com/3697. The full Molyneux quote is, "In the absence of clear science explaining different outcomes between ethnicities, Steve, what we end up

with is a bunch of social justice warriors trying to explain it using crazy things like white privilege, systemic bigotry, massive racism, and so on," at approximately 38 minutes (Hot Carnivore Coach. "Will Genius Be Genetically Engineered Stephen Hsu and Stefan Molyneux." YouTube Video, 1:32:16. May 28, 2017. https://www .youtube.com/watch?v=TivfwIhm-8w). This is no surprise; previously, Molyneux has insisted that "screaming 'racism' at people because Blacks are collectively less intelligent . . . is insane" (this quote is from a YouTube video that has since been removed for hate speech violations so the raw source is unavailable, but it is documented at the following source: "Stefan Molyneux." SPL Center, accessed June 5, 2024. https://www.splcenter.org/fighting-hate/extremist-files/individual/stefan -molyneux).

4. For Molyneux interviewing a White nationalist, see "Fighting Against Globalism." Produced by Alex Jones and Stefan Molyneux. *Freedomain Radio*. March 31, 2017. Podcast, 33:16. https://www.fdrpodcasts.com/3339/an-honest-conversation -about-race-jared-taylor-and-stßefan-molyneux. For Molyneux interviewing a conspiracy theorist, see "An Honest Conversation About Race." Produced by Jared Taylor and Stefan Molyneux. *Freedomain Radio*. July 8, 2016. Podcast, 1:18:32. https://www.fdrpodcasts.com/3637/fighting-against-globalism-alex-jones-and -stefan-molyneux.

5. The full Molyneux quote is: "If you look at the average IQ of Mexico, it tends to hover a little bit higher than the sweet spot for criminality. But, of course, in East Asian countries—as you point out, Beijing and Seoul—you get an average IQ 104, 105, 106, and so on and therefore criminality is just enormously lower."

6. Full Hsu quote: "There's a battle within our institutions of higher education and on one side are scientists often who want to preserve meritocracy and truth and other people who really have primarily political and ideological motivations and the things they want to accomplish within the institution which are not based on science or truth and that is the subtext of this battle so you basically you know if you're administrator you get caught in these kinds of uh conflicts" (see CSPI. "Diversity Hires and BLM Maoism: Steve Hsu's Cancelation Story." YouTube video, 13:42. March 23, 2022. https://www.youtube.com/watch?v=5p_Pu2Qq6mg).

7. Amy Harmon. "Why White Supremacists Are Chugging Milk (and Why Geneticists Are Alarmed)." *New York Times*, October 17, 2018. https://www.nytimes .com/2018/10/17/us/white-supremacists-science-dna.html.

Notably, lactase persistence was not a socially sensitive topic until it came to light that there were ancestral differences. No one's normative worth is tied to their ability to digest dairy, and there are not known to be systematic disparities in life outcomes between those who can and cannot. Instead, White supremacists added normative value post hoc to a trait that individuals of European ancestries happened to genetically have. That a trait like lactase persistence became social charged highlights the agility of the discourses that researchers must be sensitive to.

8. Iselin Gambert and Tobias Linné. "How the Alt-Right Uses Milk to Promote White Supremacy." *The Conversation*, April 26, 2018. http://theconversation.com /how-the-alt-right-uses-milk-to-promote-white-supremacy-94854.

9. Kevin A. Bird and Jedidiah Carlson. "Typological Thinking in Human Genomics Research Contributes to the Production and Prominence of Scientific Racism." *Frontiers in Genetics* 15 (2024). https://doi.org/10.3389/fgene.2024.1345631; Aaron Panofsky and Joan Donovan. "Genetic Ancestry Testing among White Nationalists: From Identity Repair to Citizen Science." *Social Studies of Science* 49, no. 5 (2019): 653–681. https://doi.org/10.1177/0306312719861434; Aaron Panofsky, Kushan Dasgupta, and Nicole Iturriaga. "How White Nationalists Mobilize Genetics: From Genetic Ancestry and Human Biodiversity to Counterscience and Metapolitics." *American Journal of Physical Anthropology* 175, no. 2 (2021): 387–398. https://doi.org /10.1002/ajpa.24150.

10. Jesper Pallesen et al. "Immunogenicity and Structures of a Rationally Designed Prefusion MERS-Cov Spike Antigen." *Proceedings of the National Academy of Sciences* 114, no. 25 (2017): E7348–E7357. doi: 10.1073/pnas.1707304114.

11. "727: Boulder v. Hill." This American Life, accessed June 5, 2024. https:// www.thisamericanlife.org/727/transcript. The study was eventually published in the *Proceedings of the National Academy of Sciences*. For more information, see Pallesen et al. "Immunogenicity and Structures of a Rationally Designed Prefusion MERS-CoV Spike Antigen."

12. Alexander I. Young, Stefania Benonisdottir, Molly Przeworski, and Augustine Kong. "Deconstructing the Sources of Genotype-Phenotype Associations in Humans." *Science* 365, no. 6460 (2019): 1396–1400. doi: 10.1126/science.aax 3710.

13. Melinda C. Mills et al. "Identification of 371 Genetic Variants for Age at First Sex and Birth Linked to Externalising Behaviour." *Nature Human Behaviour* 5, no. 12 (2021): 1717–1730. https://doi.org/10.1038/s41562-021-01135-3; Aysu Okbay et al. "Polygenic Prediction of Educational Attainment Within and Between Families from Genome-Wide Association Analyses in 3 Million Individuals." *Nature Genetics* 54, no. 4 (2022): 437–449. https://doi.org/10.1038/s41588-022-01016-z.

14. Benjamin W. Domingue, Daniel W. Belsky, Jason M. Fletcher, Dalton Conley, Jason D. Boardman, and Kathleen Mullan Harris. "The Social Genome of Friends and Schoolmates in the National Longitudinal Study of Adolescent to Adult Health." *Proceedings of the National Academy of Sciences* 115, no. 4 (2018): 702–707. https:// doi.org/10.1073/pnas.171180311; Abdel Abdellaoui et al. "Genetic Correlates of Social Stratification in Great Britain." *Nature Human Behaviour* 3, no. 12 (2019): 1332–1342. https://doi.org/10.1038/s41562-019-0757-5; Daniel W. Belsky et al. "Genetics and the Geography of Health, Behaviour and Attainment." *Nature Human Behaviour* 3, no. 6 (2019): 576–586; Shiro Furuya, Jihua Liu, Sun Zhongxuan,

Qiongshi Lu, and Jason M. Fletcher. "The Big (Genetic) Sort? A Research Note on Migration Patterns and Their Genetic Imprint in the United Kingdom." *Demography* 60, no. 6 (2023): 1649–1664. doi: 10.1215/00703370-11054960.

15. Augustine Kong et al. "The Nature of Nurture: Effects of Parental Genotypes." *Science* 359, no. 6374 (2018): 424–428. doi: 10.1126/science.aan6877; Timothy C. Bates et al. "The Nature of Nurture: Using a Virtual-Parent Design to Test Parenting Effects on Children's Educational Attainment in Genotyped Families." *Twin Research and Human Genetics* 21, no. 2 (2018): 73–83; Emma, Armstrong-Carter, Sam Trejo, Liam J. B. Hill, Kirsty L. Crossley, Dan Mason, and Benjamin W. Domingue. "The Earliest Origins of Genetic Nurture: The Prenatal Environment Mediates the Association Between Maternal Genetics and Child Development." *Psychological Science* 31, no. 7 (2020): 781–791. doi: 10.1177/0956797620917209.

16. Alexander I. Young et al. "Mendelian Imputation of Parental Genotypes Improves Estimates of Direct Genetic Effects." *Nature Genetics* 54 (2022): 897–905. https://doi.org/10.1038/s41588-022-01085-0.

17. Benjamin W. Domingue, Sam Trejo, Emma Armstrong-Carter, and Elliot M. Tucker-Drob. "Interactions between Poly Genic Scores and Environments: Methodological and Conceptual Challenges." *Sociological Science* 7 (2020): 46–86.

18. Silvia H. Barcellos, Leandro S. Carvalho, and Patrick Turley. "Education Can Reduce Health Differences Related to Genetic Risk of Obesity." *Proceedings of the National Academy of Sciences* 115, no. 42 (2018): E9765–E9772. https://doi.org/10.1073/pnas.1802909115; Benjamin W. Domingue, Klint Kanopka, Travis T. Mallard, Sam Trejo, and Elliot M. Tucker-Drob. "Modeling Interaction and Dispersion Effects in the Analysis of Gene-By-Environment Interaction." *Behavior Genetics* 52, no. 1 (2022): 56–64. doi: 10.1007/s10519-021-10090-8; Sam Trejo et al. "Schools as Moderators of Genetic Associations with Life Course Attainments: Evidence from the WLS and Add Health." *Sociological Science* 5 (2018): 513–540. doi: 10.15195/v5.a22; Robbee Wedow, Meghan Zacher, Brooke M. Buibregtse, Kathleen Mullan Harris, Benjamin W. Domingue, and Jason D. Boardman. "Education, Smoking, and Cohort Change: Forwarding a Multidimensional Theory of the Environmental Moderation of Genetic Effects." *American Sociological Review* 83, no. 4 (2018): 802–832; Nicholas W. Papageorge and Kevin Thom. "Genes, Education, and Labor Market Outcomes: Evidence from the Health and Retirement Study." *Journal of the European Economic Association* 18, no. 3 (2020): 1351–1399. https://doi.org/10.1093/jeea/jvz072; Victor Ronda et al. "Family Disadvantage, Gender, and the Returns to Genetic Human Capital." *Scandinavian Journal of Economics* 124, no. 2 (2022): 550–578. https://doi.org/10.1111/sjoe.12462; Rebecca Johnson, Ramina Sotoudeh, and Dalton Conley. "Polygenic Scores for Plasticity: A New Tool for Studying Gene–Environment Interplay." *Demography* 59, no. 3 (2022): 1045–1070. https://doi.org/10.1215/00703370-9957418; Ben W. Domingue, Hexuan Liu, Aysu Okbay, and

Daniel W. Belsky. "Genetic Heterogeneity in Depressive Symptoms Following the Death of a Spouse: Polygenic Score Analysis of the US Health and Retirement Study." *American Journal of Psychiatry* 174, no. 10 (2017): 963–970.

19. Pamela Herd et al. "Genes, Gender Inequality, and Educational Attainment." *American Sociological Review* 84, no. 6 (2019): 1069–1098.

20. Anushka Patil. "Gunman Targeted Black Neighborhood Shaped by Decades of Segregation." *New York Times*, May 15, 2022. https://www.nytimes.com/2022/05/14/nyregion/east-side-buffalo-shooting.html.

21. "Expressway Seen as Symbol of Racial Inequity, Health Problems." WBFO, January 15, 2018. https://www.wbfo.org/local/2018-01-15/expressway-seen-as-symbol-of-racial-inequity-health-problems. Richard Rothstein. *The Color of Law: A Forgotten History of How Our Government Segregated America*. New York: Liveright Publishing, 2017

22. Many people have told the story of Kat Massey's escapade with the Cherry Street Block club. The sociologist and author Eve Ewing told it in a 2022 episode of *This American Life*. The episode celebrated and offered insights into the lives of the ten African Americans who were killed in a Tops Supermarket in Buffalo, New York, on Saturday, May 14, 2022. For more information on this story, see Eve L. Ewing. "Name. Age. Detail: Katherine Massey." *This American Life*, August 12, 2022. https://www.thisamericanlife.org/777/name-age-detail.

23. Daphne O. Martschenko, Ben W. Domingue, Lucas J. Matthews, and Sam Trejo. "FoGS Provides a Public FAQ Repository for Social and Behavioral Genomic Discoveries." *Nature Genetics* 53, no. 9 (2021): 1272–1274.

24. In 2016, 2017, 2019, and 2021, the SSGAC held a multiday training program to introduce researchers to statistical genetic tools, methods, and concepts so that they could apply them in the social and medical sciences. Sam was an attendee of one such training, and Daphne gave a lecture on research ethics at another.

25. "Letters of Support." Letter for Stephen Hsu, accessed June 5, 2024. https://sites.google.com/view/petition-letter-stephen-hsu/additional-letters.

26. Carl Zimmer. "Years of Education Influenced by Genetic Makeup, Enormous Study Finds." *New York Times*, July 23, 2018. https://www.nytimes.com/2018/07/23/science/genes-education.html.

27. Brian Resnick. "How Scientists Are Learning to Predict Your Future with Your Genes." *Vox*, August 25, 2018. https://www.vox.com/science-and-health/2018/8/23/17527708/genetics-genome-sequencing-gwas-polygenic-risk-score.

28. Ed Yong. "An Enormous Study of the Genes Related to Staying in School." *The Atlantic* (2018). https://www.theatlantic.com/science/archive/2018/07/staying-in-school-genetics/565832.

29. Peter H. Sudmant, Tobias Rausch, Eugene J. Gardner, Robert E. Handsaker, Alexej Abyzov, John Huddleston, Yan Zhang et al. "An Integrated Map of Structural Variation in 2,504 Human Genomes." *Nature* 526, no. 7571 (2015): 75–81.

30. Jedidiah Carlson. "Combatting the Weaponization of Science." Lecture presented at Kings College London, virtually, December 2022.

31. Megan Molteni. "Buffalo Shooting Ignites a Debate Over the Role of Genetics Researchers in White Supremacist Ideology." STAT, May 23, 2022. https://www .statnews.com/2022/05/23/buffalo-shooting-ignites-debate-genetics-researchers -in-white-supremacist-ideology/.

32. Janet D. Stemwedel. "Science Must Not Be Used to Foster White Supremacy." *Scientific American*, May 24, 2022. https://www.scientificamerican.com/article /science-must-not-be-used-to-foster-white-supremacy/.

Alongside Robbee Wedow, a sociologist at Purdue University, we also wrote a piece in *Scientific American* entitled: "Scientists Must Consider the Risk of Racist Misappropriation of Research." See Robbee Wedow, Daphne O. Martschenko, and Sam Trejo. "Scientists Must Consider the Risk of Racist Misappropriation of Research." *Scientific American*, May 26, 2022. https://www-scientificamerican-com .stanford.idm.oclc.org/article/scientists-must-consider-the-risk-of-racist-misappro priation-of-research/.

33. Ashley Smart. "Field at a Crossroads: Genetics and Racial Mythmaking." *Undark*, December 12, 2022. https://race.undark.org/articles/a-field-at-a-crossroads -genetics-and-racial-mythmaking.

34. Daniela Peterka-Benton and Bond Benton. "Online Radicalization Case Study of a Mass Shooting: The Payton Gendron Manifesto." *Journal for Deradicalization*, no. 35 (2023): 1–32. https://journals.sfu.ca/jd/index.php/jd/article/view/737.

35. Jedidiah Carlson, Brenna M. Henn, Dana R. Al-Hindi, and Sohini Ramachandran. "Counter the Weaponization of Genetics Research by Extremists." *Nature* 610, no. 7932 (October 2022): 444–447. https://doi.org/10.1038/d41586-022 -03252-z.

36. Allison C. Morgan et al. "Socioeconomic Roots of Academic Faculty." *Nature Human Behavior* 6 (2022): 1625–1633. https://doi.org/10.1038/s41562-022-01425-4.

37. "Race/Ethnicity of College Faculty." National Center for Education Statistics, accessed June 5, 2024. https://nces.ed.gov/fastfacts/display.asp?id=61; "Racial and Ethnic Disparities in the United States: An Interactive Chartbook." Economic Policy Institute, June 15, 2022. https://www.epi.org/publication/disparities-chartbook/.

38. Ewen Callaway. "'The Entire Protein Universe': AI Predicts Shape of Nearly Every Known Protein." *Nature*, July 29, 2022. https://www.nature.com/articles /d41586-022-02083-2.

39. Dan Robitzki. "This Startup Says It Can Analyze Your DNA to Detect If You're Gay." *Neoscope*, October 16, 2019. https://futurism.com/neoscope/app -analyzes-dna-how-gay.

40. Christopher Pece and Gary W. Anderson. "Analysis of Federal Funding for Research Development in 2022: Basic Research." National Center for Science and Engineering Statistics, August 15, 2024. https://ncses.nsf.gov/pubs/nsf24332.

41. Michelle N. Meyer et al. "Wrestling with Social and Behavioral Genomics: Risks, Potential Benefits, and Ethical Responsibility." *Hastings Center Report* 53, no. S1 (2023): S2–S49. https://doi.org/10.1002/hast.1477.

42. Daphne O. Martschenko et al. "Wrestling with Public Input on an Ethical Analysis of Scientific Research." *Hastings Center Report* 53, no. S1 (2023): S50–S65. doi: 10.1002/hast.1478.

43. "Citizen Science." NASA, accessed June 5, 2024. https://science.nasa.gov /citizenscience. It's worth noting that although citizen science can be empowering, the citizen science research process itself can, at times, be hijacked and exploited to enact social harms. For instance, see Aaron Panofsky and Joan Donovan. "Genetic Ancestry Testing among White Nationalists: From Identity Repair to Citizen Science." *Social Studies of Science* 49, no. 5 (2019): 653–681. https://doi.org/10.1177 /0306312719861434.

44. "The Native BioData Consortium." *Native BioData Consortium*, accessed June 5, 2024. https://nativebio.org/about-us/.

45. Susan Sheridan, Suzanne Schrandt, Laura Forsythe, Tandrea S. Hilliard, and Kathryn A. Paez. "The PCORI Engagement Rubric: Promising Practices for Partnering in Research." *Annals of Family Medicine* 15, no. 2 (2017): 165–170. https://doi .org/10.1370/afm.2042.

46. Healthy Flint Research Coordinating Center, accessed June 5, 2024. https:// www.hfrcc.org/. Flint Center for Health Equity Solutions, accessed June 5, 2024. https://flintcenter.org/.

47. Shobita Parthasarathy. "Can Innovation Serve the Public Good?" *Boston Review*, July 6, 2023. https://www.bostonreview.net/articles/can-innovation-serve-the -public-good/.

48. Alon Zivony, Rasha Kardosh, Liadh Timmins, and Niv Reggev. "Ten Simple Rules for Socially Responsible Science." *PLOS* (2023). https://doi.org/10.1371 /journal.pcbi.1010954.

49. Genevieve L. Wojcik. "Eugenics Is on the Rise Again: Human Geneticists Must Take a Stand." *Nature* 641, no. 8061 (May 2025): 37–38. https://doi.org/10 .1038/d41586-025-01297-4. See also Jedidiah Carlson, Brenna M. Henn, Dana R. Al-Hindi, and Sohini Ramachandran. "Counter the Weaponization of Genetics Research by Extremists." *Nature* 610, no. 7932 (October 2022): 444–47. https://doi .org/10.1038/d41586-022-03252-z.

6. DNA and Social Inequality

1. For more details on Sam's decision to donate, see Sam Trejo. "Op-Ed: I Donated My Kidney to a Stranger—and More of Us Should." *Los Angeles Times*, January 19, 2020. https://www.latimes.com/opinion/story/2020-01-19/kidney-trans plants-donors-shortage.

2. Katherine P. Harden. *The Genetic Lottery: Why DNA Matters for Social Equality.* Princeton, NJ: Princeton University Press, 2021.

3. Michael Barbaro. "Why Is the Pandemic Killing So Many Black Americans?" Produced by Adizah Eghan et al. *The Daily.* May 20, 2020. Podcast, 32:16. https://www.nytimes.com/2020/05/20/podcasts/the-daily/black-death-rate-coronavirus.html.

4. Lindsay M. Monte and Daniel J. Perez-Lopez. "How the Pandemic Affected Black and White Households." US Census Bureau, July 21, 2021. https://www.census.gov/library/stories/2021/07/how-pandemic-affected-black-and-white-households.html.

5. Jason Laughlin and John Duchneskie. "COVID Has Killed More than 5,000 Philadelphians. These Neighborhoods Lost the Most." *Philadelphia Inquirer,* April 11, 2022. https://www.inquirer.com/health/coronavirus/philadelphia-deaths-5000-covid-pandemic-poverty-race-age-northeast-20220411.html.

6. "COVID-19 Population Characteristics." SF.gov. Accessed April 28, 2024. https://sf.gov/data/covid-19-population-characteristics.

7. CDC. Cases, Data, and Surveillance. Centers for Disease Control and Prevention. https://web.archive.org/web/20220121032952/https://www.cdc.gov/coronavirus/2019-ncov/covid-data/investigations-discovery/hospitalization-death-by-race-ethnicity.html (2020).

8. "Uninsured Rates for the Nonelderly by Race/Ethnicity." The Kaiser Family Foundation. Accessed April 28, 2024. https://www.kff.org/uninsured/state-indicator/nonelderly-uninsured-rate-by-raceethnicity/.

9. Richard V. Reeves and Faith Smith. "Black and Hispanic Americans at Higher Risk of Hypertension, Diabetes, Obesity: Time to Fix Our Broken Food System." Brookings, August 7, 2020. https://www.brookings.edu/blog/up-front/2020/08/07/black-and-hispanic-americans-at-higher-risk-of-hypertension-diabetes-obesity-time-to-fix-our-broken-food-system/.

10. Anthony B. Atkinson. *Measuring Poverty around the World.* Princeton, NJ: Princeton University Press, 2019.

11. Raj Chetty et al. "The Association Between Income and Life Expectancy in the United States, 2001–2014." *JAMA* 315, no. 16 (2016): 1750–1766. doi: 10.1001/jama.2016.4226.

12. Elizabeth L. Tung et al. "Race/Ethnicity and Geographic Access to Urban Trauma Care." *JAMA Network Open* 2, no. 3 (2019): e190138. doi: 10.1001/jamanetworkopen.2019.0138.

13. Some thinkers, like the eighteenth-century Swiss philosopher Jean-Jacques Rousseau, distinguish between so-called natural inequality—such as variation in inborn physical abilities which are thought to be inherent and unchanging—and other forms of inequality, such as those inequalities created by social processes and institutions. On this view, one may believe that we have a moral imperative to reduce

various social inequalities but *not* natural inequalities. However, we see this conceptual distinction as inherently fraught; as the Millie and Marcia Biggs example in chapter 2 highlights, the impact of a person's natural characteristics (e.g., their DNA) can often operate through long causal pathways that include social and environmental forces. Moreover, many natural inequalities, like poor eyesight, are readily ameliorated via social intervention (eyeglasses in the case of poor eyesight). For these reasons, in this book, we do not distinguish between natural and other forms of inequality when considering our collective moral commitments to build a more socially equal world.

14. Charles Tilly. *Durable Inequality*. Berkeley and Los Angeles: University of California Press, 1999. Structural inequality is an umbrella term. It may include, for example, structural racism (sometimes also referred to as institutional or systemic racism) or structural ableism, and it can be applied to specific contexts, such as education or the economy, or more broadly to society. The specific structural categories that are advantaged or disadvantaged by a given social structure may vary across place and time.

15. Equality of opportunity and outcome are two interrelated concepts. Equality of opportunity typically refers to everyone being given the same chance of achieving some outcome (e.g., applying to college). Equality of opportunity should ideally result in equality of outcome, but in order for that to happen, certain individuals will need to receive different resources or supports depending upon their circumstances (e.g., need-based financial support for college application fees). Researchers and laypeople sometimes use the word "equity" to describe this process of allocating resources and supports depending upon someone's circumstances.

16. Bill Dedman. "Fulton's Michael Lomax: 'If I can't get a loan, what black person can?'" *Atlanta Journal-Constitution*, May 1, 1988. https://www.ajc.com/news/fulton-michael-lomax-can-get-loan-what-black-person-can/vhnKLXz2H1zJnO86j38s4I/.

17. Dedman, Bill. "The Color of Money: Home Mortgage Lending Practices Discriminate Against Blacks." *Atlanta Journal and Atlanta Constitution*, 1988.

18. Dedman, "The Color of Money."

19. Richard Rothstein. *The Color of Law: A Forgotten History of How Our Government Segregated America*. New York: Liveright Publishing, 2017.

20. Daniel Aaronson, Daniel Hartley, and Bhashkar Mazumder. "The Effects of the 1930s HOLC 'Redlining' Maps." *American Economic Journal: Economic Policy* 13, no. 4 (2021): 355–392. doi: 10.1257/pol.20190414.

21. Lincoln Quillian, John J. Lee, and Brandon Honoré. "Racial Discrimination in the U.S. Housing and Mortgage Lending Markets: A Quantitative Review of Trends, 1976–2016." *Race and Social Problems* 12 (2020): 13–28.

22. Some scholars argue that the total effects of redlining are far more expansive than merely denying specific individuals access to credit and home ownership. In

particular, redlining emerged as a key part of a broader transition in the United States to a new economy of private housing based on systematic speculation of property values. This system served to economically legitimize a racist valuation scheme—the idea that a property's price could be straightforwardly forecasted simply based on proximity to Black Americans—which in turn led to a broad devaluation of Black property and communities. As Historian Keeanga-Yamahtta Taylor puts it: "People talk about the free market as this racially neutral, color-blind space within which the invisible hand of supply and demand dictates what does and doesn't happen. But that's so incredibly naïve. The market is us. The market is a reflection of our values. And when it comes to property, race is at the very center" (see Sean Illing. "The Sordid History of Housing Discrimination in America." *Vox*, May 5, 2020. https://www.vox.com/identities/2019/12/4/20953282/racism-housing-discrimination-keeanga-yamahtta-taylor.)

23. Neil Bhutta et al. "Disparities in Wealth by Race and Ethnicity in the 2019 Survey of Consumer Finances." Board of Governors of the Federal Reserve System, September 28, 2020. https://www.federalreserve.gov/econres/notes/feds-notes/disparities-in-wealth-by-race-and-ethnicity-in-the-2019-survey-of-consumer-finances-20200928.html; Dalton Conley. *Being Black, Living in the Red: Race, Wealth, and Social Policy in America, 10th Anniversary Edition, with a New Afterword.* Berkeley and Los Angeles: University of California Press, 2009.

24. Harden, for instance, argues in *The Genetic Lottery* that "inequalities that are due to factors over which people have no control . . . cannot be said to [be] deserve[d]." (Perhaps the key thesis of her book is combining this idea with the premise that "none of us deserves his or her genetics.") To those who distinguish between undeserved and deserved inequality, the mere existence in differences in well-being between individuals does not necessarily imply the existence of social inequality, as a portion (or even all) of these differences could be deserved. Nonetheless, for any given philosophical conceptualization of deservingness, the key points presented in this chapter still hold. That is, for those who believe that we should focus on reducing undeserved differences in well-being, there exists a tension between addressing average between-group differences in undeserved well-being and addressing other within-group variation in undeserved well-being. Moreover, we argue that the two key sources of variation in well-being used as examples in this chapter—(i) differences in access to credit due to one's race and (ii) variation in educational opportunity resulting from one's precise date of birth—are both plausibly undeserved.

25. Bhashkar Mazumder. "Family and Community Influences on Health and Socioeconomic Status: Sibling Correlations Over the Life Course." *B.E. Journal of Economic Analysis & Policy* 11, no. 3 (2011): 2876. doi: 10.2202/1935-1682.2876; Bhashkar Mazumder. "Sibling Similarities and Economic Inequality in the US." *Journal of Population Economics* 21, no. 3 (2008): 685–701. https://www.jstor.org/stable

/40344699; Timothy J. Halliday and Bhashkar Mazumder. "An Analysis of Sibling Correlations in Health Using Latent Variable Models." *Health Economics* 26, no. 12 (2017): e108–e125. doi: 10.1002/hec.3483.

26. Sandra E. Black, Paul J. Devereux, and Kjell G. Salvanes. "Too Young to Leave the Nest? The Effects of School Starting Age." *Review of Economics and Statistics* 93, no. 2 (2011): 455–467. https://www.jstor.org/stable/23015947; Philip J. Cook and Songman Kang. "Birthdays, Schooling, and Crime: Regression-Discontinuity Analysis of School Performance, Delinquency, Dropout, and Crime Initiation." *American Economic Journal: Applied Economics* 8, no. 1 (2016): 33–57. doi: 10.1257/app.20140323.

27. David Deming and Susan Dynarski. "The Lengthening of Childhood." *Journal of Economic Perspectives* 22, no. 3 (2008): 71–92. doi: 10.1257/jep.22.3.71.

28. The apparent misconception that entering school later is beneficial is likely the result of so-called apples to oranges comparisons. For example, if you compared Max and Sam's third grade tests scores, Sam's would likely be higher. However, this is not because he has benefited from being relatively older than his peers, but instead because he would be nine and a half when the test was administered, while Max would be just nine. If you instead compared scores on tests that were administered when each brother was exactly nine, Max is likely to have the advantage given that he entered school earlier.

29. Black et al. "Too Young to Leave the Nest?"; Cook and Kang. "Birthdays, Schooling, and Crime"; Elizabeth U. Cascio and Diane W. Schanzenbach. "First in the Class? Age and the Education Production Function." *Education Finance and Policy* 11, no. 3 (2016): 225–250.

30. Todd E. Elder. "The Importance of Relative Standards in ADHD Diagnoses: Evidence Based on Exact Birth Dates." *Journal of Health Economics* 29, no. 5 (2010): 641–656. doi: 10.1016/j.jhealeco.2010.06.003.

31. The importance of inequality for individuals caused by school-starting age cutoffs compared to inequality resulting from redlining is a function of the differences in well-being created by each process. So, while one might *count* all sources of inequality, it may be that in practice certain sources are much more impactful (and therefore potentially better targets for policies intended to reduce inequality).

32. While some studies have reported associations between season of birth and race (see Kasey S. Buckles and Daniel M. Hungerman. "Season of Birth and Later Outcomes: Old Questions, New Answers." *Review of Economics and Statistics* 95, no. 3 (2013): 711–724. doi: 10.1162/REST_a_00314), the magnitude of these relationships is substantively small and the vast majority of the variation in birth date remains unexplained by race. Moreover, due to quasi-random variation in school entry thresholds across place and time, differences in season of birth across racial group do not necessarily entail average differences in school-starting entry across groups.

33. In order for some causal factor, be it the age a child enters kindergarten or whether they experience lead poisoning, to produce disparities across structural categories, it must be unequally distributed across those categories. Lead paint has been outlawed in the United States since the 1970s and can be readily contained by a few fresh coats of lead-free paint. This means that poorer children, who are more likely to live in older homes with decaying walls, are the most likely to ingest lead. Lead exposure, then, is among the many factors that contribute to structural inequality. However, the effects of school entry age do not produce structural inequality because, unlike lead exposure, they are not unequally patterned across groups to a meaningful degree.

34. Dalton Conley and Benjamin Domingue. "The Bell Curve Revisited: Testing Controversial Hypotheses with Molecular Genetic Data." *Sociological Science* 3 (2016): 520–539.

35. Charles Murray. "Genetics Will Revolutionize Social Science." *Wall Street Journal*, January 27, 2020. https://www.wsj.com/articles/genetics-will-revolutionize-social-science-11580169106.

36. Patrick Turley, Michelle Meyer, and Daniel Benjamin. "Genetic Scoring Presents Opportunity, Peril." *Wall Street Journal*, February 3, 2020. https://www.wsj.com/articles/genetic-scoring-presents-opportunity-peril-11580762369.

37. Importantly, the question of whether DNA-based measures of genetic risk can ever, in principle, inform efforts to reduce social inequality (which is our focus in the current chapter) is distinct from the question of whether existing polygenic scores can do so. Indeed, many of the critiques of *The Genetic Lottery* centered on the claim that Harden inadequately highlighted the conceptual limitations (which we discuss in chapter 2) and practical limitations (see chapters 7 and 8) of current polygenic scores. For more information, see Graham Coop and Molly Przeworski. "Lottery, Luck, or Legacy: A Review of 'The Genetic Lottery: Why DNA Matters for Social Equality.'" *Evolution* 76 (2022): 846–853. https://doi.org/10.1111/evo.14449.

38. Daphne O. Martschenko. "Social Equality in an Alternate World." *Hastings Center Report* 51, no. 6 (2021): 54–55. doi: 10.1002/hast.1307.

39. Brenna M. Henn, Emily Klancher Merchant, Anne O'Connor, and Tina Rulli. "Why DNA Is No Key to Social Equality: On Kathryn Paige Harden's 'The Genetic Lottery.'" *Los Angeles Review of Books*, September 21, 2021. https://lareviewofbooks.org/article/why-dna-is-no-key-to-social-equality-on-kathryn-paige-hardens-the-genetic-lottery/.

40. Aysu Okbay et al. "Polygenic Prediction of Educational Attainment within and between Families from Genome-Wide Association Analyses in 3 Million Individuals." *Nature Genetics* 54 (2022): 437–449. https://doi.org/10.1038/s41588-022-01016-z; Sam Trejo and Klint Kanopka. "Using the Phenotype Differences Model to Identify Genetic Effects in Samples of Partially Genotyped Sibling Pairs." *Proceedings of the National Academy of Sciences* 121, no. 49 (2024): e2405725121.

41. Harden, *The Genetic Lottery*.

42. However, recent work suggests that certain genomic measures, known as genetic similarity proportions, may be useful for measuring differences in the experience of racialization by others among members of the same racial group. These proportions closely correspond to the global information provided by popular genetic ancestry tests. Luyin Zhang and Sam Trejo. "DNA, Self-Reported Ancestry, and Social Scientific Inquiry." *SocArXiv* (2025).

43. This statistic refers to the difference between the 75th and 25th percentiles of the income distribution. For more information, see Erica Hanushek, Jacob Light, Paul E. Peterson, Laura M. Talpey, and Ludger Woessmann. "Long-Run Trends in the U.S. SES-Achievement Gap." *Education Finance Policy* 17, no. 4 (2022): 608–640.

44. While there exists no scientific evidence to support the idea that race and class disparities, the structural inequalities that we focus on in this book, are the inevitable result of DNA differences across groups, the case of other structural inequalities is sometimes more complex; many physical and psychological impairments are partly genetically caused, and these conditions can play a role in the production of health and social disparities between those with and without disabilities. Notably, while certain disabilities may be influenced by DNA, the impact of a given disability on a person's well-being can often result from ableist aspects of our social arrangement. Simply the fact that DNA influences one's likelihood of being a member of a given structural category (e.g., disabled) does not indicate that the use of DNA is required to help reduce that type of structural inequality (e.g., well-being disparities between people with and without disabilities).

45. Ruha Benjamin. *Viral Justice: How We Grow the World We Want*. Princeton, NJ: Princeton University Press, 2022.

46. Matthew Desmond. *Poverty, by America*. New York: Penguin, 2023.

47. Matthew J. Salganik, Ian Lunberg, Alexander T. Kindel, and Sara McLanahon. "Measuring the Predictability of Life Outcomes with a Scientific Mass Collaboration." *Proceedings of the National Academy of Sciences* 117, no. 15 (2020): 8398–8403. https://doi.org/10.1073/pnas.1915006117.

7. Polygenic Prediction at the Fertility Clinic

1. Andrew Niccol, director. *Gattaca*. Columbia Pictures, 1997.

2. Niccol. *Gattaca*.

3. The He Lab. "About Lulu and Nana: Twin Girls Born Healthy After Gene Surgery as Single-Cell Embryos." YouTube video, 4:43. November 25, 2018. https://www.youtube.com/watch?v=th0vnOmFltc&ab_channel=TheHeLab.

4. Dennis Normile. "CRISPR Bombshell: Chinese Researcher Claims to Have Created Gene-Edited Twins." *Science*, November 26, 2018. https://www.science.org/content/article/crispr-bombshell-chinese-researcher-claims-have-created-gene-edited-twins.

5. *Human Genome Editing: Science, Ethics, and Governance*. Washington, DC: National Academies Press, 2017; "On Human Gene Editing: International Summit Statement." National Academies of Science, Engineering, and Medicine, December 3, 2015. https://www.nationalacademies.org/news/2015/12/on-human-gene-editing-international-summit-statement.

6. Eric S. Lander et al. "Adopt a Moratorium on Heritable Genome Editing." *Nature* 567 (2019): 165–168. doi: https://doi.org/10.1038/d41586-019-00726-5.

7. "Human Genome Editing: Recommendations." World Health Organization, July 12, 2021. https://www.who.int/publications/i/item/9789240030381.

8. Yong Xie, Shaohua Zhan, Wei Ge, and Peifu Tang. "The Potential Risks of C-C Chemokine Receptor 5-Edited Babies in Bone Development." *Bone Research* 7 (2019): 4. doi: 10.1038/s41413-019-0044-0.

9. Miou Zhou et al. "CCR5 Is a Suppressor for Cortical Plasticity and Hippocampal Learning And Memory." *eLife* 5 (2016): e20985. doi: 10.7554/eLife.20985; Antonio Regalado. "China's CRISPR Twins Might Have Had Their Brains Inadvertently Enhanced." *MIT Technology Review*, February 21, 2019. https://www.technologyreview.com/2019/02/21/137309/the-crispr-twins-had-their-brains-altered/.

10. Dana Goodyear. "The Transformative, Alarming Power of Gene Editing." *The New Yorker*, September 2, 2023. https://www.newyorker.com/magazine/2023/09/11/the-transformative-alarming-power-of-gene-editing.

11. Goodyear. "The Transformative, Alarming Power of Gene Editing."

12. "The US' First Test Tube Baby." PBS, accessed June 2, 2024. https://www.pbs.org/wgbh/americanexperience/features/babies-americas-first/.

13. Carey Goldberg. "Just Another Girl, Unlike Any Other." *New York Times*, October 27, 1999. https://www.nytimes.com/1999/10/27/us/just-another-girl-unlike-any-other.html.

14. Ciara Nugent. "What It Was Like to Grow Up as the World's First 'Test-Tube Baby.'" *Time*, July 25, 2018. https://time.com/5344145/louise-brown-test-tube-baby/.

15. "New Survey Analysis on Morality of Abortion, Stem Cell Research and IVF." Pew Research Center, August 15, 2013. https://www.pewresearch.org/religion/2013/08/15/new-survey-analysis-on-morality-of-abortion-stem-cell-research-and-ivf/; Gabriel Borelli. "Americans Overwhelmingly Say Access to IVF Is a Good Thing." Pew Research Center, May 13, 2024. https://www.pewresearch.org/short-reads/2024/05/13/americans-overwhelmingly-say-access-to-ivf-is-a-good-thing/.

16. Alander Rocha. "Alabama Passed a New IVF Law. But Questions Remain." *Alabama Reflector*, March 11, 2024. https://alabamareflector.com/2024/03/11

/alabama-passed-a-new-ivf-law-but-questions-remain/; Praveena Somasundaram. "Alabama Governor Signs IVF Bill Giving Patients, Providers Legal Cover." *Washington Post*, March 6, 2024. https://www.washingtonpost.com/nation/2024/03/06/alabama-governor-signs-ivf-bill/.

17. The policy framework introduced in this book is guided by elements of what some scholars call the "precautionary principle." The precautionary principle is more frequently applied to assessments of environmental risk. Although this principle is the subject of debate (e.g., in what contexts is it justifiable to invoke the precautionary principle), our policy framework captures its general spirit by seeking to balance innovation with risk reduction in situations of technological and scientific uncertainty. For more on the precautionary principle, see Carl F. Cranor. "Toward Understanding Aspects of the Precautionary Principle." *Journal of Medicine and Philosophy: A Forum for Bioethics and Philosophy of Medicine* 29, no. 3 (2004): 259–279. https://doi.org/10.1080/03605310490500491.

18. Antonio Regalado. "Eugenics 2.0: We're at the Dawn of Choosing Embryos by Health, Height, and More." *MIT Technology Review*, November 1, 2017. https://www.technologyreview.com/2017/11/01/105176/eugenics-20-were-at-the-dawn-of-choosing-embryos-by-health-height-and-more/.

19. Antonio Regalado. "America's First IVF Baby Is Pitching a Way to Pick the DNA of Your Kids." *MIT Technology Review*, April 26, 2023. https://www.technologyreview.com/2023/04/26/1071533/embryo-prediction-marketer-future-job-titles/.

20. "Frequently Asked Questions." *Genomic Prediction*, accessed June 2, 2024. https://web.archive.org/web/20200610080736/https://genomicprediction.com/faqs/.

21. Rachel Pells. "Genetic Screening Now Lets Parents Pick the Healthiest Embryos." *Wired*, July 5, 2022. https://www.wired.com/story/genetic-screening-ivf-healthiest-embryos/; Rachel Pells. "Aurea Is First Baby in the World to Be 'Selected' After Genetic Screen." *Daily Mail*, October 17, 2022. https://www.dailymail.co.uk/health/article-11324851/Aurea-baby-world-selected-genetically-screened.html; Pells. "Genetic Screening Now Lets Parents Pick the Healthiest Embryos."

22. Genomic Prediction. "Rank Ordering Embryos for Transfer, Utilizing PGT-P: Patient and Clinician Perspectives." YouTube video, 1:34:11. May 20, 2021. https://www.youtube.com/watch?v=HE5ADe7BgdM.

23. Teresa Carey. "Genetic Testing Promises Parents the Ability to Select Healthiest Embryo." WHYY, June 10, 2022. https://whyy.org/segments/startup-offers-genetic-testing-that-promises-to-predict-healthiest-embryo/.

24. Julianna LeMieux. "Polygenic Risk Scores and Genomic Prediction: Q&A with Stephen Hsu." *Genetic Engineering & Biotechnology News*, April 1, 2019. https://www.genengnews.com/insights/polygenic-risk-scores-and-genomic-prediction-qa-with-stephen-hsu/.

25. Remy A. Furrer et al. "Public Attitudes, Interests, and Concerns Regarding Polygenic Embryo Screening." *JAMA* 7, no. 5 (2024): e2410832. doi:10.1001/jamanetworkopen.2024.10832.

26. For more information, see Nathan Slotnick. "Genetic Testing for Autism, Intellectual Disability, and Other Neurodevelopmental Disorders." *Orchid*, June 13, 2023. https://guides.orchidhealth.com/post/genetic-testing-for-autism-intellectual-disability-and-other-neurodevelopmental-disorders. In December 2023, Orchid Health announced its plans to offer polygenic embryo selection for psychiatric conditions, such as schizophrenia. The company was promptly criticized by the Psychiatric Genomics Consortium, an international group of researchers responsible for producing the most accurate polygenic score for schizophrenia. The consortium was concerned that their discoveries were not being used to "improve the lives of people with mental illness" but instead to "stop them from being born" (see Carrie Arnold. "Genetics Group Slams Company for Using Its Data to Screen Embryos' Genomes." *Science*, December 15, 2023. https://www.science.org/content/article/genetics-group-slams-company-using-its-data-screen-embryos-genomes). The consortium further claimed that Orchid's use of its schizophrenia polygenic score violated the data sharing agreement through which they had made their findings publicly available to other researchers. An Orchid spokesperson later said that their polygenic score did not use information from the consortium's studies, but Orchid would not reveal which study's results their polygenic score are based on.

27. Hannah Devlin, Tom Burgis, David Pegg, and Jason Wilson. "US Startup Charging Couples to Screen Embryos for IQ." *The Guardian*, October 18, 2024. https://www.theguardian.com/science/2024/oct/18/us-startup-charging-couples-to-screen-embryos-for-iq.

28. "IVF by the Numbers." *Penn Medicine*, March 14, 2018. https://www.pennmedicine.org/updates/blogs/fertility-blog/2018/march/ivf-by-the-numbers.

29. Genomic Prediction. "Rank Ordering Embryos for Transfer, Utilizing PGT-P: Patient and Clinician Perspectives." YouTube video, 1:34:11. May 20, 2021. https://www.youtube.com/watch?v=HE5ADe7BgdM.

30. D. Barlevy, I. Cenolli, T. Campbell, R. Furrer, M. Mukherjee, K. Kostick-Quenet, S. Carmi, T. Lencz, G. Lazaro-Munoz, and S. Pereira. "Patient Interest in and Clinician Reservations on Polygenic Embryo Screening: A Qualitative Study of Stakeholder Perspectives." *Journal of Assisted Reproduction and Genetics* 41, no. 5 (2024): 1221–1231. https://doi.org/10.1101/2023.10.12.23296961.

31. Rémy A. Furrer, Dorit Barlevy, Stacey Pereira, Shai Carmi, Todd Lencz, and Gabriel Lázaro-Muñoz. "Public Attitudes, Interests, and Concerns Regarding Polygenic Embryo Screening." *JAMA Network Open* 7, no. 5 (2024): e2410832–e2410832. In addition, interview studies suggest that clinicians may be more concerned about uses of the technology than their patients.

32. Michelle N. Meyer, Tammy Tan, Daniel J. Benjamin, David Laibson, and Patrick Turley. "Public Views on Polygenic Screening of Embryos." *Science* 379, no. 6632 (2023): 541–543.

33. Simone Zhang, Rebecca A. Johnson, John Novembre, Edward Freeland, and Dalton Conley. "Public Attitudes Toward Genetic Risk Scoring in Medicine and Beyond." *Social Science & Medicine* 274 (2021): 113796. doi: 10.1016/j.socscimed.2021.113796.

34. Frank B. Gilbreth and Ernestine G. Carey. *Cheaper by the Dozen.* Thomas Y. Crowell Co. 1948.

35. Authors' calculations of the average male and female height in the United States stem from weighted Wave IV data of the National Longitudinal Study of Adolescent to Adult Health, 69.9 inches for males and 64.4 inches for females.

36. To reduce the incidence of twin and high-order multiple gestations, which carry greater risks than singleton gestations, the American Society for Reproductive Medicine generally recommends that clinicians transfer only a single embryo at a time. However, under certain rare circumstances (e.g., advanced maternal age), clinicians are permitted to transfer more than one embryo. For more information, see "Guidance on the Limits to the Number of Embryos to Transfer: A Committee Opinion." American Society for Reproductive Medicine. https://www.asrm.org /practice-guidance/practice-committee-documents/guidance-on-the-limits-to-the -number-of-embryos-to-transfer-a---committee-opinion-2021/.

37. Edgar Dahl et al. "Preconception Sex Selection Demand and Preferences in the United States." *Fertility and Sterility* 85, no. 2 (2006): 468–473. doi: 10.1016/j.fertnstert.2005.07.1320.

38. Jeff Wang and Mark V. Sauer. "In Vitro Fertilization (IVF): A Review of 3 Decades of Clinical Innovation and Technological Advancement." *Therapeutics and Clinical Risk Management* 2, no. 4 (2006): 355–364. doi: 10.2147/tcrm.2006.2.4.355.

39. The number of eggs retrieved during a single round of IVF ranges from the low single digits to over 30. Only a fraction of eggs that are retrieved each round will eventually become viable embryos for implantation. For more information, see Leigh A. Humphries et al. "Is Younger Better? Donor Age Less than 25 Does Not Predict More Favorable Outcomes after In Vitro Fertilization." *Journal of Assisted Reproduction and Genetics* 36, no. 8 (2019): 1631–1637. doi: 10.1007/s10815-019-01494-x; and Åsa Magnusson, Karin Källen, Ann Thurin-Kjellberg, and Christina Bergh. "The Number of Oocytes Retrieved during IVF: A Balance between Efficacy and Safety." *Human Reproduction* 33, no. 1 (2018): 58–64. https://doi.org/10.1093 /humrep/dex334.

40. Felix R. Day, Ken K. Ong, and John R. B. Perry. "Elucidating the Genetic Basis of Social Interaction and Isolation." *Nature Communications* 9 (2018): 2457. https://doi.org/10.1038/s41467-018-04930-1.

41. Thousands of Americans have elected to undergo limb-lengthening surgeries to increase their height by a similar amount (see Uwa Ede-Osifo and Lauren Wilson. "Leg-lengthening Surgery Is Gaining Popularity Among Men Seeking to Be Taller, Doctors Say." NBC News, April 23, 2023. https://www.nbcnews.com/health/mens -health/leg-lengthening-surgery-gains-popularity-men-seeking-taller-rcna79819; and Ashish Mittal et al. "Trends and Practices in Limb Lengthening: An 11-Year US Database Study." *Strategies in Trauma and Limb Reconstruction* 18, no. 1 (2023): 22–31. 10.5005/jp-journals-10080-1574). These procedures involve cutting into the thigh bones in each leg and inserting rods inside them, which are slowly lengthened via remote control. In addition to being painful, these procedures can cost as much as $100,000. Of note, throughout the process of writing this book, we have noticed there is considerable variation in what degree of expected gain any given person considers to be meaningful; while some folks may think that a gain of a couple inches in height is worth it, others may see it as trivial.

42. In the statistical genetics literature, the term liability is typically used (rather than risk).

43. This useful trick is made possible by the fact that both height and heart disease risk form a bell-curve distribution (which, in statistics, is also known as the normal distribution).

44. Patrick Turley, Michelle N. Meyer, Nancy Wang, David Cesarini, Evelynn Hammonds, Alicia R. Martin, Benjamin M. Neale, et al. "Problems with Using Polygenic Scores to Select Embryos." *New England Journal of Medicine* 385, no. 1 (2021): 78–86.

45. As a reminder, whenever we use terms like European ancestries or Indigenous American ancestries, we are deploying the concept of geographic ancestry categories (introduced in chapter 3).

46. Yi Ding, Kangcheng Hou, Ziqi Xu, Aditya Pimplaskar, Ella Petter, Kristin Boulier, Florian Privé, Bjarni J. Vilhjálmsson, Loes M. Olde Loohuis, and Bogdan Pasaniuc. "Polygenic Scoring Accuracy Varies Across the Genetic Ancestry Continuum." *Nature* 618, no. 7966 (2023): 774–781.

47. Sam Trejo, Gloria Yeomans-Maldonado, and Brian Jacob. "The Effects of the Flint Water Crisis on the Educational Outcomes of School-Age Children." *Science Advances* 10, no. 11 (2024): eadk4737. https://www.science.org/doi/10.1126/sciadv .adk4737.

48. The overall effects of a focal individual inheriting a set of genetic variants can be separated into (i) the initial effects of the focal individual's DNA on their own outcomes, and (ii) the subsequent effects of the change in the focal individual's outcomes on the average outcomes of their broader community. By and large, the genomic studies that have produced existing polygenic scores have, thus far, explored the former but not the latter—perhaps partly a result of the need to prioritize sample

size and statistical power. Thus, it is possible that a set of genetic variants that posi-
tively affect an individual's outcome have no effect (or, in certain cases, even negative
effects) on the average outcome among individuals in a broader community.

49. Thorstein Veblen. *The Theory of the Leisure Class*. New York: Macmillan, 1899;
Fred Hirsch. *Social Limits to Growth*. Cambridge, MA: Harvard University Press, 1976.

50. Garrett Hardin. "The Tragedy of the Commons." *Science* 162, no. 3859
(1968): 1243–1248. doi:10.1126/science.162.3859.1243.

51. Marissa Conrad. "How Much Does IVF Cost?" *Forbes*, August 14, 2023.
https://www.forbes.com/health/womens-health/how-much-does-ivf-cost/; "What
to Know About I.V.F." *New York Times*, April 18, 2020. https://www.nytimes.com
/article/ivf-treatment.html; Andrew D.A.C. Smith, Kate Tilling, Scott M. Nelson,
and Debbie A. Lawlor. "Live-Birth Rate Associated with Repeat In Vitro Fertilization
Treatment Cycles." *JAMA* 314, no. 24 (2015): 2654–2662. doi: 10.1001/jama.2015
.17296.

52. Phil Galewitz. "Even When IVF Is Covered by Insurance, High Bills, Sur-
prises and Hassles Abound." NPR, May 4, 2022. https://www.npr.org/sections
/health-shots/2022/05/04/1095589987/ivf-insurance-bills.

53. Alexander G. Ioannidis et al. "Paths and Timings of the Peopling of Polynesia
Inferred from Genomic Networks." *Nature* 597 (2021): 522–526. https://doi.org
/10.1038/s41586-021-03902-8.

54. Nancy P. Gordon, Teresa Y. Lin, Jyoti Rau, and Joan C. Lo. "Aggregation of
Asian-American Subgroups Masks Meaningful Differences in Health and Health
Risks Among Asian Ethnicities: An Electronic Health Record Based Cohort Study."
BMC Public Health 19, no. 1 (2019): 1551. doi: 10.1186/s12889-019-7683-3.

55. "Island Health." Centers for Disease Control and Prevention, accessed June 3,
2024. https://web.archive.org/web/20230127155836/https://www.cdc.gov
/chronicdisease/resources/publications/factsheets/island-health.htm. (Note that,
as of this writing in early 2025 [at the beginning of President Trump's second term
in office], the research and data listed on the CDC website and the websites of other
federal agencies are in flux.)

56. This statistic was calculated using the difference between the graduations
rates of economically disadvantaged (80–81%) and all students (86-87%) using data
from the National Center for Education Statistics (NCES) (see Table 219.46. Public
High School 4-Year Adjusted Cohort Graduation Rate (ACGR), by Selected Student
Characteristics and State: 2010–11 through 2018–19." National Center for Education
Statistics. Accessed May 28, 2024, https://nces.ed.gov/programs/digest/d20/tables
/dt20_219.46.asp; and Table 219.46. Public High School 4-Year Adjusted Cohort
Graduation Rate (ACGR), by Selected Student Characteristics And State: 2010–11
through 2019–20. National Center for Education Statistics. Accessed May 28, 2024,
https://nces.ed.gov/programs/digest/d21/tables/dt21_219.46.asp), leveraging the
fact that approximately 50% of US students qualify as economically disadvantaged

(authors' calculations; Erin M. Fahle et al. "Stanford Education Data Archive: Technical Documentation: Version 2.0." December 4, 2017. https://cepa.stanford.edu/sites/default/files/SEDA_documentation_v20b.pdf).

57. Another potential concern is decreased human genetic and, in turn, phenotypic diversity. Normally, each new generation born represents a roll of the dice, a random reshuffling of the parent's DNA that introduces variability and uniqueness into our population. However, the more that polygenic embryo selection is used, the more this roll is made with loaded dice—rather than having the same chance of rolling 1, 2, 3, 4, 5, and 6, these dice land on 5 or 6 nearly every time. Over time, if polygenic embryo selection is continually applied for many traits, humans may become increasingly similar to one another; however, if (or how quickly) this could occur is still uncertain. Decreased genetic and phenotypic diversity would not only threaten to limit the richness of our social world, but also may adversely impact our species' resilience against the evolution of new diseases. Interestingly, in the Genomic Prediction marketing video in which Rafal appeared, he discusses the possibility that polygenic embryo selection may reduce human genetic diversity but says, "here we're talking about the diversity where some people get sick and that's not the kind of diversity we want to maintain." Genomic Prediction. "Rank Ordering Embryos for Transfer, Utilizing PGT-P: Patient and Clinician Perspectives." YouTube video, 1:34:11. May 20, 2021. https://www.youtube.com/watch?v=HE5ADe7BgdM.

58. Celeste M. Condit. "Laypeople Are Strategic Essentialists, Not Genetic Essentialists." *Looking for the Psychosocial Impact of Genomic Information* 49, no. 1 (2019): S27–S37. https://doi.org/10.1002/hast.1014.

59. Andrew M. McIntosh et al. "Genetic and Environmental Risk for Chronic Pain and the Contribution of Risk Variants for Major Depressive Disorder: A Family-Based Mixed-Model Analysis." *PLOS Medicine* 13, no. 8 (2016): e1002090. doi: 10.1371/journal.pmed.1002090; Keira J. A. Johnston et al. "Genome-wide Association Study of Multisite Chronic Pain in UK Biobank." *PLOS Genetics* 15, no. 6 (2019): e1008164. doi: 10.1371/journal.pgen.1008164.

60. Rhys Blakely. "Patients Warned Against American Clinics Offering 'Unethical' Embryo Screening." *The Times*, accessed June 2, 2024. https://www.thetimes.co.uk/article/patients-warned-against-american-clinics-offering-unethical-embryo-screening-zkb76zqq7.

61. Human Fertilisation and Embryology Authority, accessed June 2, 2024. https://www.hfea.gov.uk/.

62. Sarah Kliff. "F.D.A. Wants to Regulate Prenatal Tests That Have Misled Parents." *New York Times*, June 27, 2023. https://www.nytimes.com/2023/06/27/upshot/prenatal-testing-misleading-fda.html.

63. "FDA Takes Action Aimed at Helping to Ensure the Safety and Effectiveness of Laboratory Developed Tests." U.S. Food & Drug Administration, April 29, 2024. https://www.fda.gov/news-events/press-announcements/fda-takes-action-aimed

-helping-ensure-safety-and-effectiveness-laboratory-developed-tests; "ACLA Challenge to FDA's Final Rule Regulating Laboratory Developed Testing Services as Medical Devices." ACLA, May 29, 2024. https://www.acla.com/acla-challenge-to-fdas
-final-rule-regulating-laboratory-developed-testing-services-as-medical-devices/.

64. Note that universal reproductive care in the lead-up to and during pregnancy
is distinct from universal healthcare. Universal healthcare would require a much
bigger shift in how the US healthcare system is organized.

65. Lisa Dive and Ainsley J. Newson. "Reproductive Carrier Screening: Respond-
ing to the Eugenics Critique." *Journal of Medical Ethics* 48, no. 12 (2022): 1060–1067.

66. Michelle Jokisch Polo. "Infertility Patients Fear Abortion Bans Could Affect
Access to IVF Treatment." NPR, July 21, 2022. https://www.npr.org/sections/health
-shots/2022/07/21/1112127457/infertility-patients-fear-abortion-bans-could
-affect-access-to-ivf-treatment.

67. Philip Reilly. "Commentary: The Federal 'Prenatally and Postnatally Diagnosed Conditions Awareness Act.'" *Prenatal Diagnosis* 29, no. 9 (2009): 829–832.
doi: 10.1002/pd.2304.

68. Antonio Regalado. "The World's First Gattaca Baby Tests Are Finally Here."
MIT Technology Review, November 8, 2019. https://www.technologyreview.com
/2019/11/08/132018/polygenic-score-ivf-embryo-dna-tests-genomic-prediction
-gattaca/.

69. The first quote comes from Gattaca Genomics, accessed June 2, 2024.
https://gattacagenomics.com/about-us.html; the second and fourth quotes come
from Orchid Health, accessed June 2, 2024. https://web.archive.org/web/202409
27031246/https://www.orchidhealth.com/; the third quote comes from Lifeview,
Powered by Genomic Prediction, accessed June 2, 2024. https://web.archive.org
/web/20230901235008/https://www.lifeview.com/.

70. Noor Siddiqui, X post, April 23, 2025 at 8:12AM PT, https://x.com/noor
siddiqui/status/1915061139819676122. Siddiqui's post on X (formerly Twitter)
isn't the first time that she has suggested parents ought to pursue polygenic embryo
selection. In a 2021 podcast, she staked her claim around the idea of "earned" and
"unearned" luck. Siddiqui explained that getting hit by a car would be a form of
unearned back luck, saying: "that's totally out of your control." She went on to say
that "going base jumping constantly" and then breaking your leg was a form of
"earned bad luck"; "You kind of exposed yourself to higher risk there," she continued.
For Siddiqui, not utilizing polygenic embryo selection is akin to opening someone
up to earned bad luck when polygenic embryo selection could reduce a potential
child's exposure to potential risks (see Theral Timpson. "Orchid Health Is 1st in the
World to Offer Whole Genome Couple's Report." Mendelspod, April 6, 2021.
https://www.mendelspod.com/p/orchid-health-is-1st-in-the-world-b49). In a different video on X, Siddiqui said: "It's going to become insane not to screen for these
things" (see Anna Louie Sussman. "Opinion | Should Human Life Be Optimized?"

New York Times, April 1, 2025, sec. Opinion. https://www.nytimes.com/interactive /2025/04/01/opinion/ivf-gene-selection-fertility.html). Also see Jessica Winter. "How Much Should You Know About Your Child Before He's Born?" *The New Yorker*, April 21, 2025. https://www.newyorker.com/magazine/2025/04/28/how -much-should-you-know-about-your-child-before-hes-born.

71. For more on epidemiological, clinical, and ethical considerations regarding polygenic embryo selection, see Antonio Capalbo, Guido de Wert, Heidi Mertes, Liraz Klausner, Edith Coonen, Francesca Spinella, Hilde Van de Velde, et al. "Screening Embryos for Polygenic Disease Risk: A Review of Epidemiological, Clinical, and Ethical Considerations." *Human Reproduction Update* (2024): dmae012. https://doi .org/10.1093/humupd/dmae012.

8. Polygenic Prediction in the Hands of Consumers and Institutions

1. "About Us." Pronatalist.org, accessed June 2, 2024, https://pronatalist.org /aboutus/; Jenny Kleeman. "America's Premier Pronatalists on Having 'Tons of Kids' to Save the World: 'There are Going to Be Countries of Old People Starving to Death.'" *The Guardian*, May 25, 2024. https://www.theguardian.com/lifeandstyle /article/2024/may/25/american-pronatalists-malcolm-and-simone-collins.

2. Dean Spears. "All of the Predictions Agree on One Thing: Humanity Peaks Soon." *New York Times*, September 18, 2023. https://www.nytimes.com/interactive /2023/09/18/opinion/human-population-global-growth.html; Dean Spears and Michael Geruso. *After the Spike: Population, Progress, and the Case for People*. New York: Simon & Schuster, 2025.

3. "The Art of Media Baiting: Inside the Tactics of the Pronatalist Movement." Produced by Based Camp: Simone & Malcolm Collins. *Aporia*. January 18, 2024. Podcast, 32:40. https://basedcamppodcast.substack.com/p/the-art-of-media -baiting-inside-the

4. "Reversing the Fertility Collapse." Written by Malcolm Collins. *Aporia*. March 13, 2024. https://www.aporiamagazine.com/p/reversing-the-fertility -collapse.

5. Carey Goldberg. "The Pandora's Box of Embryo Testing Is Officially Open." *Bloomberg*, May 26, 2022. https://www.bloomberg.com/news/features/2022-05-26 /dna-testing-for-embryos-promises-to-predict-genetic-diseases.

6. Ara Mahdawi. "'Hipster Eugenics': Why Is the Media Cosying Up to People Who Want to Build a Super Race?" *The Guardian*, April 21, 2023. https://www.the guardian.com/lifeandstyle/2023/apr/20/pro-natalism-babies-global-population -genetics.

7. Kleeman. "America's Premier Pronatalists on Having 'Tons of Kids' to Save the World."

8. "Direct-to-Consumer Genetic Testing FAQ for Healthcare Professionals." National Human Genome Research Institute, accessed June 2, 2024. https://www
.genome.gov/For-Health-Professionals/Provider-Genomics-Education-Resources
/Healthcare-Provider-Direct-to-Consumer-Genetic-Testing-FAQ; Imran Khan.
"Spit Parties: Genetic Testing Becomes a Social Activity." *The Guardian,* September 18, 2008. https://www.theguardian.com/science/blog/2008/sep/18/genetic
.testing.

9. These fees assume genotyping or DNA sequencing has been completed previously. Genotyping or low-coverage sequencing typically costs consumers around
$100.

10. Susan Ipaktchian. "Medical School to Offer Course that Gives Students Option of Studying Their Own Genotype Data." Stanford Medicine: News Center,
June 7, 2010. https://med.stanford.edu/news/all-news/2010/06/medical-school-to
-offer-course-that-gives-students-option-of-studying-their-own-genotype-data
.html; Sam has even taught one such course himself! Although we do not cover genetic ancestry testing in detail, these tests are popular among consumers and have
had a profound impact on social life. For instance, long-lost family members find
each other by uploading their DNA to relative-finding platforms, while law enforcement access their data to solve previously unsolvable crimes (see Jocelyn Kaiser. "We
Will Find You: DNA Search Used to Nab Golden State Killer Can Home in on About
60% of White Americans." *Science,* October 11, 2018. https://www.science.org
/content/article/we-will-find-you-dna-search-used-nab-golden-state-killer-can
-home-about-60-white). Some individuals receive their genetic ancestry test results
and begin questioning, or even changing, their racial identities (see Ruth Padawer.
"Sigrid Johnson Was Black. A DNA Test Said She Wasn't." *New York Times,* November 19, 2018. https://www.nytimes.com/2018/11/19/magazine/dna-test-black
-family.html).

11. "Hunter Farmer DNA Report" Genomelink, accessed June 2, 2024. https://
genomelink.io/product/hunter-farmer-ancestry-dna-report; "Is Breast Size Genetic?"
Genomelink, accessed June 2, 2024. https://genomelink.io/traits/breast-size.

12. "Couples Report: Discover Deep Love Compatibility with DNA Romance's
Advanced Couples Compatibility Report," May 16, 2025. https://web.archive.org
/web/20250516053901/https://dnaromance.com/couples-report/.

13. Bowen Hu et al. "Genome-wide Association Study Reveals Sex-Specific Genetic Architecture of Facial Attractiveness." *PLOS Genetics* 15, no. 4: e1007973.
https://doi.org/10.1371/journal.pgen.1007973. Note, this study used data from the
Wisconsin Longitudinal Study (WLS), which acquired the high school yearbooks
from most schools for the graduating class that comprised the WLS cohort. Each
yearbook photo was then rated by six men and six women using a photo-labeled 11-point rating scale, with end points labeled as "not at all attractive" (=1) and "extremely
attractive" (=11). For more information, see "Wisconsin Longitudinal Study (WLS)

[graduates, siblings, and spouses]: 1957–2020 Version 14.03. [machine-readable data file]." University of Wisconsin–Madison, accessed June 4, 2024. https://researchers .wls.wisc.edu/documentation/.

14. "DNA Influences Facial Attractiveness." Genomelink, accessed June 2, 2024. https://web.archive.org/web/20230605003443/https://genomelink.io/traits /facial-attractiveness. A person with a low score (tenth percentile) and a person with a high score (ninetieth percentile) would likely only have a minuscule difference in how others actually rate their attractiveness: about *a tenth* of one point on a classic 1 to 10 "attractiveness" scale. This conservative estimate assumes the higher of the two sex-specific estimates of SNP heritability reported in Bowen Hu et al. "Genome-wide Association Study Reveals Sex-Specific Genetic Architecture of Facial Attractiveness." Using the formula described in Ronald De Vlaming et al. "Meta-GWAS Accuracy and Power (MetaGAP) Calculator Shows That Hiding Heritability Is Partially Due to Imperfect Genetic Correlations Across Studies." *PLoS Genetics* 13, no. 1 (2017): e1006495, a sample size of 4,383, and the default polygenicity parameters, the resulting polygenic score is estimated to have an out-of-sample R^2 equal to 0.00134. The within-sex standard deviation of attractiveness (measured using an 11-point scale) was calculated by the authors using the raw Wisconsin Longitudinal Study data to be approximately 1.26 points.

15. Early on, 23andMe also offered tests for social and behavioral traits like intelligence, reading ability, breastfeeding, and muscle performance. After having to discontinue marketing their personal genetics services in 2013, the company never brought these tests back. There does not seem to be a legal reason for this decision.

16. "FDA Warning Letter to 23andMe." Casewatch, November 27, 2023. https:// quackwatch.org/cases/fdawarning/prod/fda-warning-letters-about-products-2013 /23andme/.

17. Bill Sutton. "Overview of Regulatory Requirements: Medical Devices—Transcript." US Food and Drug Administration, January 4, 2018. https://www.fda .gov/training-and-continuing-education/cdrh-learn/overview-regulatory-require ments-medical-devices-transcript. The 23andMe health-related genetic tests were labeled Class II medical devices. Class II medical devices present a moderate risk of harm (e.g., powered wheelchairs). For more information, see Madison K. Kilbride, Susan M. Domchek, and Angela R. Bradbury. "How Should Patients and Providers Interpret the US Food and Drug Administration's Regulatory Language for Direct-to-Consumer Genetic Tests?" *Journal of Clinical Oncology* 31, no. 28 (2019): 2514–2517. doi: 10.1200/JCO.18.01418.

18. Jacob S. Sherkow, Jin K. Park, and Christine Y. Lu. "Regulating Direct-to-Consumer Polygenic Risk Scores." *JAMA* 330, no. 8 (2023): 691–692. doi:10.1001/ jama.2023.12262.

19. Sherkow et al. "Regulating Direct-to-Consumer Polygenic Risk Scores"; Jin K. Park and Christine Y. Lu. "Polygenic Scores in the Direct-to-Consumer

Setting: Challenges and Opportunities for a New Era in Consumer Genetic Testing." *Journal of Personalized Medicine* 13, no. 4 (2023): 573. https://doi.org/10.3390/jpm13040573.

20. Sherkow et al. "Regulating Direct-to-Consumer Polygenic Risk Scores"; Park and Lu. "Polygenic Scores in the Direct-to-Consumer Setting."

21. Daniel J. Benjamin et al. "The Genetic Architecture of Economic and Political Preferences." *PNAS* 109, no. 21 (2012): 8026–8031 (2012). https://doi.org/10.1073/pnas.1120666109.

22. "Are Your Political Views Affected by DNA?" Genomelink, accessed June 2, 2024. https://genomelink.io/product/political-dna-test-report.

23. Sherkow et al. "Regulating Direct-to-Consumer Polygenic Risk Scores."

24. SelfDecode, accessed June 2, 2024. https://selfdecode.com/.

25. Michelle Fernandes Martins, Logan T. Murry, Lisel Telford, and Frank Moriarty. "Direct-to-Consumer Genetic Testing: An Updated Systematic Review of Healthcare Professionals' Knowledge and Views, and Ethical and Legal Concerns." *European Journal of Human Genetics* 30, no. 12 (2022): 1331–1343. doi: 10.1038/s41431-022-01205-8.

26. George Annas and Sherman Elias. "23andMe and the FDA." *New England Journal of Medicine* 370, no. 11 (2014): 985–988. https://www.nejm.org/doi/full/10.1056/NEJMp1316367.

27. David J. Kaufman, Juli M. Bollinger, Rachel L. Dvoskin, and Joan A. Scott. "Risky Business: Risk Perception and the Use of Medical Services among Customers of DTC Personal Genetic Testing." *Journal of Genetic Counseling* 21 (2012): 413–422. https://doi.org/10.1007/s10897-012-9483-0; Serena Oliveri et al. "What People Really Change after Genetic Testing (GT) Performed in Private Labs: Results from an Italian Study." *European Journal of Human Genetics* 30 (2022): 62–72. https://doi.org/10.1038/s41431-021-00879-w.

28. Tia Moscarello, Brittney Murray, Chloe M. Reuter, and Erin Demo. "Direct-to-Consumer Raw Genetic Data and Third-Party Interpretation Services: More Burden than Bargain?" *Genetics in Medicine* 21, no. 3 (2019): 539–541. https://doi.org/10.1038/s41436-018-0097-2.

29. Lucas J. Matthews, Matthew S. Lebowitz, Ruth Ottman, and Paul S. Appelbaum. "Pygmalion in the Genes? On the Potentially Negative Impacts of Polygenic Scores for Educational Attainment." *Social Psychology of Education* 24 (2021): 789–808. doi:10.1007/s11218-021-09632-z; Lucas J. Matthews, Zhijun Zhang, and Daphne O. Martschenko. "Schoolhouse Risk: Can We Mitigate the Polygenic Pygmalion Effect?" *Acta Pyschologica* 248 (2024): 104403. https://doi.org/10.1016/j.actpsy.2024.104403; Matthew S. Lebowitz and Woo-Kyoung Ahn. "Blue Genes? Understanding and Mitigating Negative Consequences of Personalized Information about Genetic Risk for Depression." *Journal of Genetic Counseling* 27, no. 1 (2018): 204–216. doi: 10.1007/s10897-017-0140-5.

30. Timothy Caulfield and Amy L. McGuire. "Direct-to-Consumer Genetic Testing: Perceptions, Problems, and Policy Responses." *Annual Review of Medicine* 63 (2012): 23–33. doi: 10.1146/annurev-med-062110-123753.

31. "Are Your Political Views Affected by DNA?" Genomelink, accessed June 2, 2024. https://genomelink.io/product/political-dna-test-report.

32. "ACLA Challenges FDA's Final Rules to Regulate Laboratory Developed Testing Services as Medical Devices." *American Clinical Laboratory Association*, May 29, 2024. https://www.acla.com/acla-challenges-fdas-final-rule-to-regulate -laboratory-developed-testing-services-as-medical-devices/.

33. Martins et al. "Direct-to-Consumer Genetic Testing: An Updated Systematic Review of Healthcare Professionals' Knowledge and Views, and Ethical and Legal Concerns."

34. "President Bush Signs Law Banning Internet Gambling." PBS NewsHour, October 16, 2006. https://www.pbs.org/newshour/show/president-bush-signs-law -banning-internet-gambling.

35. "Genetic Counselors and Their Roles at 23andMe." 23andMe Blog (blog), November 11, 2022. https://blog.23andme.com/articles/genetic-counselors-and -their-roles-at-23andme; Madison K. Kilbride, Susan M. Domchek, and Angela R. Bradbury. "How Should Patients and Providers Interpret the US Food and Drug Administration's Regulatory Language for Direct-to-Consumer Genetic Tests?" *Journal of Clinical Oncology* 31, no. 28 (2019): 2514–2517. doi: 10.1200/JCO.18 .01418.

36. Kaiser. "We Will Find You: DNA Search Used to Nab Golden State Killer Can Home in on About 60% of White Americans."

37. Yaniv Erlich and Dina Zielinski. "Anatomy of the 23andMe Fall and Implications for Consumer Genomics." *Nature Biotechnology*, May 29, 2025, 1–3. https://doi .org/10.1038/s41587-025-02683-z.

38. Rylee Kirk. "23andMe Customers Did Not Expect Their DNA Data Would Be Sold, Lawsuit Claims." *New York Times*, June 10, 2025, sec. Business. https://www .nytimes.com/2025/06/10/business/23andme-data-lawsuit.html.

39. Jeffrey Osei et al. "Polygenic Risk Scores in Clinical Practice? Still Making the Case." Centers for Disease Control and Prevention, July 5, 2022. https://blogs.cdc .gov/genomics/2022/07/05/polygenic-risk-scores/.

40. Early Check, accessed June 2, 2024. https://earlycheck.org/; Armstrong, Ben. "Polygenic Score Pilot for Heart Disease Begins." Genomics Education Programme (blog), January 28, 2022. https://www.genomicseducation.hee.nhs.uk /blog/polygenic-score-pilot-for-heart-disease-begins/.

41. Amit Sud et al. "Realistic Expectations Are Key to Realising the Benefits of Polygenic Scores." *The BMJ* 380 (2023): e073149. doi: 10.1136/bmj-2022-073149.

42. Kathryn Asbury and Robert Plomin. *G Is for Genes: The Impact of Genetics on Education and Achievement*. Hoboken, NJ: Wiley-Blackwell, 2013.

43. Kathryn Asbury, Tom McBride, and Kaili Rimfeld. "Genetics and Early Intervention: Exploring Ethical and Policy Questions." Early Intervention Foundation, August 11, 2021. https://www.eif.org.uk/report/genetics-and-early -intervention-exploring-ethical-and-policy-questions.

44. "Gruen Lab Projects: New Haven Lexinome Project." Yale School of Medicine, accessed June 2, 2024. https://medicine.yale.edu/lab/gruen/projects/. Elite gifted-education programs like Duke University's Talent Identification Program and the Johns Hopkins Center for Talented Youth have been approached by researchers hoping to genotype "gifted" children and identify variants associated with high intelligence (see Elaine Tuttle Hansen, Stuart Gluck, and Amy L. Shelton. "Obligations and Concerns of an Organization Like the Center for Talented Youth." *Hastings Center Report* 45, no. 5 (2015): S66–72. doi: 10.1002/hast.502). In addition, while calling for education reform, some academics have suggested that improving school experiences "might prevent genetic influences on crime from unfolding" (see J. Wertz et al. "Genetics and Crime: Integrating New Genomic Discoveries into Psychological Research About Antisocial Behavior." *Psychological Science* 29, no. 5 (2018): 791–803. doi: 10.1177/0956797617744542).

45. M. Sabatello et al. "Nature vs. Nurture in Precision Education: Insights of Parents and the Public." *American Journal of Bioethics* 13, no. 2 (2022): 79–88. doi: 10.1080/23294515.2021.1983666.

46. Sian Griffiths. "Genetic Study Aims to Help Poor, Bright Children Succeed." *Sunday Times*, July 14, 2019. https://www.thetimes.co.uk/article/genetic-study-aims -to-help-poor-bright-children-succeed-bjfmq58v3.

47. American Society of Human Genetics Board of Directors and American College of Medical Genetics Board of Directors. "Points to Consider: Ethical, Legal, and Psychosocial Implications of Genetic Testing in Children and Adolescents." *American Journal of Human Genetics* 97, no. 5 (2015): 1233–1241.

48. Mandy Kendrick. "China Children's Camp Tests DNA—*Gattaca* Becomes More Than Science Fiction." *Scientific American* blog, August 5, 2009. https://blogs .scientificamerican.com/news-blog/chinese-childrens-camp-tests-dnagat-2009-08 -05/.

49. Juliet Macur. "Born to Run? Little Ones Get Test for Sports Gene." *New York Times*, November 30, 2008. https://www.nytimes.com/2008/11/30/sports /30genetics.html.

50. "Let Your DNA Journey Begin." *Circle DNA*. https://web.archive.org/web /20240105102419/https://circledna.com/en/.

51. "Our Story." Collins Institute, accessed June 2, 2024. https://collinsinstitute .org/ourstory/.

52. "Why Does the Legacy School System Work the Way It Does?" Collins Institute, accessed June 2, 2024. https://collinsinstitute.org/why/.

53. For a more detailed discussion on recommendations for using polygenic scores in clinic, see A. Abu-El-Haija et al. The Clinical Application of Polygenic Risk Scores: A Points to Consider Statement of the American College of Medical Genetics and Genomics (ACMG). *Genetics in Medicine* 25 (2023). https://www.gimjournal.org/article/S1098-3600(23)00816-X/fulltext.

54. Michelle N. Meyer Nicholas W. Papageorge, Erik Parens, Alan Regenberg, Jeremy Sugarman, and Kevin Thom. "Potential Corporate Uses of Polygenic Indexes: Starting a Conversation about the Associated Ethics and Policy Issues." *Perspective* 111, no. 5 (2024): 833–840. https://doi.org/10.1016/j.ajhg.2024.03.010.

55. At the time GINA was codified, genetic discrimination was not a rampant issue. But supporters of precision health feared that the mere risk of genetic discrimination would stymie their efforts to advance their agenda (for more see James Tabery. *Tyranny of the Gene: Personalized Medicine and Its Threat to Public Health.* First edition. New York: Alfred A. Knopf, 2023). Although limited in its scope today given the pace of technological progress, GINA is an example of a law meant to anticipate harms and mitigate against them. Note also that GINA does not apply to employers who have fewer than 15 employees.

56. Kristen V. Brown. "Genetic Discrimination Is Coming for Us All." *The Atlantic*, November 12, 2024. https://www.theatlantic.com/health/archive/2024/11/dna-genetic-discrimination-insurance-privacy/680626/.

57. Reps. Sprowls, Williams, et al. "HB 1189—Genetic Information for Insurance Purposes." Florida Senate, accessed June 2, 2024. https://www.flsenate.gov/Committees/BillSummaries/2020/html/2232.

58. Americans with Disabilities Act. US Department of Labor, accessed 2, 2024. https://web.archive.org/web/20250112164127/www.dol.gov/general/topic/disability/ada. Importantly, the ADA does not apply to risk factors for a given disability (e.g., a high polygenic score for the trait in question) but instead only to actual manifestations of that disability.

59. Though the *Students for Fair Admissions v. Harvard* Supreme Court ruling deemed affirmative action in college admissions to be unconstitutional, there remain other domains in which affirmative action remains permissible. For instance, laws still permit the use of affirmative action in the workplace; it is still legal for employers to employ affirmative action to enhance workplace diversity.

60. Elisa Jillson. "Selling Genetic Testing Kits? Read On." Federal Trade Commission, March 21, 2019. https://www.ftc.gov/business-guidance/blog/2019/03/selling-genetic-testing-kits-read. The agency's list of recommendations also includes the following: (1) Explain third-party disclosures clearly; (2) Disclose key information clearly and conspicuously; (3) Explain who can see what profile information—and let users know about important changes; and (4) Consider one-stop shopping for expunging genetic information. For further reading on DTC

testing in the United States, see Mary A. Majumder, Christi J. Guerrini, and Amy L. McGuire. "Direct-to-Consumer Genetic Testing: Value and Risk." *Annual Review of Medicine* 72, no. 1 (2021): 151–166. https://doi.org/10.1146/annurev-med-070119 -114727.

61. Robert Chapman et al. "New Literacy Challenge for the Twenty-First Century: Genetic Knowledge Is Poor Even Among Well-Educated." *Journal of Community Genetics* 10, no. 1 (2019): 73–84. doi: 10.1007/s12687-018 -0363-7.

62. Brian M. Donovan. "Ending Genetic Essentialism through Genetics Education." *Human Genetics and Genomics Advances* 3, no. 1 (2022): 10058. doi: 10.1016/j. xhgg.2021.100058.

63. Roxanne Parrott and Rachel A. Smith. "Defining Genes Using 'Blueprint' Versus 'Instruction' Metaphors: Effects for Genetic Determinism, Response Efficacy, and Perceived Control." *Health Communication* 29, no. 2 (2014): 137–146. doi: 10.1 080/10410236.2012.729181; John Lynch, Jennifer Bevan, Paul Achter, Tina Harris, and Celeste M. Condit. "A Preliminary Study of How Multiple Exposures to Messages about Genetics Impact on Lay Attitudes Towards Racial and Genetic Discrimination." *New Genetics and Society* 27, no. 1 (2008): 43–56. https://doi.org/10.1080 /14636770701843634.

64. Brian M. Donovan et al. "Genomics Literacy Matters: Supporting the Development of Genomics Literacy through Genetics Education Could Reduce the Prevalence of Genetic Essentialism." *Journal of Research in Science Teaching* 58, no. 4 (2021): 520–550. https://doi.org/10.1002/tea.21670; Roxanne Parrott and Rachel A. Smith. "Defining Genes Using 'Blueprint' Versus 'Instruction' Metaphors: Effects for Genetic Determinism, Response Efficacy, and Perceived Control." *Health Communication* 29, no. 2 (2014): 137–146. doi: 10.1080/10410236.2012 .729181.

9. The Future

1. Perhaps something akin to the National Academies process of bringing together diverse academic perspectives to grapple with issues such as the use of population descriptors like race, ethnicity, and ancestry in genomics research that we briefly discussed in chapter 3. For more information, see "Use of Race Ethnicity and Ancestry as Population Descriptors in Genomics Research." National Academies of Science, Engineering, and Medicine, accessed June 2, 2024. https:// www.nationalacademies.org/our-work/use-of-race-ethnicity-and-ancestry-as -population-descriptors-in-genomics-research. Also see *Rethinking Race and Ethnicity in Biomedical Research*. Washington, DC: National Academic Press, 2024.

Technical Appendix

1. See Ehud Karavani et al. "Screening Human Embryos for Polygenic Traits Has Limited Utility." *Cell* 149, no. 6 (2019): 1424–1435. doi: 10.1016/j.cell.2019.10.033.

2. Patrick Turley, Michelle N. Meyer, Nancy Wang, David Cesarini, Evelynn Hammonds, Alicia R. Martin, Benjamin M. Neale, et al. "Problems with Using Polygenic Scores to Select Embryos." *New England Journal of Medicine* 385, no. 1 (2021): 78–86.

See Alicia R. Martin et al. "Clinical Use of Current Polygenic Scores May Exacerbate Health Disparities." *Nature Genetics* 51 (2019): 585–591. https://doi.org/10.1038/s41588-019-0379-x.

INDEX

Note: Page numbers in italic type indicate figures or illustrations.

abortion, 160

academia, 99

ADHD, 120, 154, 176, 192

adversarial collaboration, xii–xiii, 199n2

affirmative action, 180–81, 243n59

All of Us research program, 149

Alison, Archibald, 20

Altman, Sam, 160, 162

American Civil Liberties Union (ACLU), 38

American Enterprise Institute, 33

American Society for Reproductive Medicine, 232n36

American Society of Human Genetics, 176

Americans with Disabilities Act (1990), 180

ancestry: defined, 209n19; Family Tree of humanity and, 44–45, 52, 55–64; geographic categorization of, 40, 56, 58–64, 213n49; Miss America beauty pageant and, 75; polygenic-informed screenings and, 148–50, *150*, 153–54, 158, 178; race in relation to, 14, 40–41, 45, 47, 52, 55, 58–64, 186. *See also* family trees

Ancestry.com, 40, 63, 163, 182

ancestry tests. *See* direct-to-consumer (DTC) genetic testing

anti-miscegenation laws, 36–40

Asian Americans, 210n31. *See also* Chinese immigrants

Atlanta Journal, The, 113

Atlantic (magazine), 98

Atlantic Mortgage, 112

attractiveness. *See* facial/physical attractiveness

Bazile, Leon M., 39

Becker, Joel, 201n7

Benjamin, Ruha, 68, 124

"Better Baby" contests, 74–75

Biggs, Marcia, 17–18, *18*, 32, 52, 55, 137, 185

Biggs, Michael, 17

Biggs, Millie, 17–18, *18*, 32, 52, 55, 137, 185

Black, Hugo, 39–40, 208n7

Black Americans: COVID's effect on, 109; intelligence claims about, 54, 98; plantation owners' attitudes toward, 50–51, 211n34; reflections on race by, 58; stereotypes of, 48, 102; structural disadvantages faced by, 112–14

black box prediction, 178–79

Braun, Mike, 40

Brown, Louise, 136

Buck, Carrie, 68–72, *73*, 74, 214n4